Principles
of Solidification

WILEY SERIES ON
THE SCIENCE AND TECHNOLOGY OF MATERIALS
Advisory Editors:
J. H. Hollomon, J. E. Burke, B. Chalmers, R. L. Sproull, A. T. Tobolsky

FRACTURE OF STRUCTURAL MATERIALS
  A. S. Tetelman and A. J. McEvily, Jr.
ORGANIC SEMICONDUCTORS
  F. Gutmann and L. E. Lyons
INTERMETALLIC COMPOUNDS
  J. H. Westbrook, editor
THE PHYSICAL PRINCIPLES OF MAGNETISM
  Allan H. Morrish
FRICTON AND WEAR OF MATERIALS
  Ernest Rabinowicz
HANDBOOK OF ELECTRON BEAM WELDING
  R. Bakish and S. S. White
PHYSICS OF MAGNETISM
  Sōshin Chikazumi
PHYSICS OF III-V COMPOUNDS
  Otfried Madelung (translation by D. Meyerhofer)
PRINCIPLES OF SOLIDIFICATION
  Bruce Chalmers
APPLIED SUPERCONDUCTIVITY
  Vernon L. Newhouse
THE MECHANICAL PROPERTIES OF MATTER
  A. H. Cottrell
THE ART AND SCIENCE OF GROWING CRYSTALS
  J. J. Gilman, editor
SELECTED VALUES OF THERMODYNAMIC PROPERTIES OF METALS AND ALLOYS
  Ralph Hultgren, Raymond L. Orr, Philip D. Anderson and Kenneth K. Kelly
PROCESSES OF CREEP AND FATIGUE IN METALS
  A. J. Kennedy
COLUMBIUM AND TANTALUM
  Frank T. Sisco and Edward Epremian, editors
MECHANICAL PROPERTIES OF METALS
  D. McLean
THE METALLURGY OF WELDING
  D. Séférian (translation by E. Bishop)
THERMODYNAMICS OF SOLIDS
  Richard A. Swalin
TRANSMISSION ELECTRON MICROSCOPY OF METALS
  Gareth Thomas
PLASTICITY AND CREEP OF METALS
  J. D. Lubahn and R. P. Felgar
INTRODUCTION TO CERAMICS
  W. D. Kingery
PROPERTIES AND STRUCTURE OF POLYMERS
  Arthur V. Tobolsky
PHYSICAL METALLURGY
  Bruce Chalmers
FERRITES
  J. Smit and H. P. J. Wijn
ZONE MELTING, SECOND EDITION
  William G. Pfann
THE METALLURGY OF VANADIUM
  William Rostoker

Principles
of Solidification

by Bruce Chalmers

GORDON McKAY PROFESSOR OF METALLURGY
DIVISION OF ENGINEERING AND APPLIED PHYSICS
HARVARD UNIVERSITY

John Wiley & Sons, Inc.
New York · London · Sydney

Preface

The purpose of this book is to provide a critical review of the state of knowledge and understanding of the process of solidification, defined for this purpose as the discontinuous change of state from liquid to crystalline solid. Much of the recent progress in this field has come from work on metallic systems, but many of the principles that have been uncovered relate equally to metallic and to nonmetallic materials.

The present seems an appropriate time for such a review because the very rapid progress during the last fifteen years in elucidating the various separate aspects of solidification has not been matched by the application of this knowledge to the problems encountered in industry. One reason may be that the results of recent research are not accessible to or assimilable by those most closely concerned with the solution of industrial problems or the creation of technological innovations.

Those engaged in research on solidification have become aware of two different points of view; they might be described as that of the physicist and that of the metallurgist, or materials scientist, respectively. The physicist tends to divide the whole subject into discrete problems, each of which can, in principle, be solved without reference to the others. The materials scientist, or metallurgist, on the other hand, recognizes that the whole problem is more than the sum of its component parts because of the interactions between them. The physicist usually aims at developing a theory of the influence of one variable on another, and expects to test it quantitatively by experiment; his product is a physical model which can be described and examined mathematically. The metallurgist, on the other hand, must often be content to understand qualitatively how several variables can be manipulated simultaneously to produce a desired result; he recognizes that it will seldom be possible to predict from theory the quantitative solution to his problem. It is my opinion, however, that progress in the industrial application of solidification processes will be greatly accelerated by the proper exploitation of the results of research. The empirical method, based on well-understood principles combined with experience, may often be the best way of extrapolating from past achievement, as in many other fields in which a new science is wedded to an established industrial art.

This book is intended to provide an understanding of the physical processes that relate to solidification and to show how these processes combine to produce the phenomena observed in practical situations. It should not be inferred that our understanding is complete. Many problems, some of considerable importance, have been identified but have not yet been solved; attention is drawn to these challenging subjects for further research.

I have resisted the temptation to give an exhaustive, and therefore unselective, account of the past work on solidification; it is my belief that perspective and understanding, and perhaps a few doubts, will contribute more to the solution of problems and the synthesis of ideas than access to an encyclopedic account of all that has been done. Therefore, I have been selective in regard to the ideas developed and the references given. The discerning reader may detect a tendency to concentrate attention on work with which the author has been associated, either directly or indirectly; but I hope that the result is nevertheless a well-balanced account of the present state of our understanding.

The physical basis of solidification can conveniently be studied on the atomic, the microscopic, and the macroscopic scales. At the atomic level, we are concerned with the atomic processes by which a crystal grows, or because of which it does not grow; and we are equally interested in the nucleation process by which a new crystal is formed. Events at the microscopic level are dominated by the local flow of heat and by the diffusion of solutes within the solidifying liquid; and on a macroscopic scale the flow of liquid metal into a mold and the extraction of heat into and through the mold are the dominant processes. Each of these topics is considered in turn in Chapter 1 through 7. Chapter 8 is a discussion of the structure (in its broadest sense) of cast metals, and of how the various aspects of structure are influenced or controlled by the physical parameters that govern the actual process of solidification. An appendix on the preparation of single crystals by solidification is included in recognition of the growing importance of single crystals in research and for industrial application, and because many of the principles discussed in the earlier chapters can be applied with advantage to the solution of problems that arise in crystal growing.

Cambridge, Mass.   Bruce Chalmers
April, 1964

Acknowledgments

I should like to express my gratitude to the many students, colleagues, and other friends with whom I have studied solidification during the last twenty years. Much of the point of view and many of the ideas expressed in this book have come from them; my indebtedness to them has perhaps made me less responsive than I should have been to much excellent work in other places. It would be impossible to name everyone to whom I would like to convey thanks, but I would be ungracious not to mention Professor W. C. Winegard, of the University of Toronto; Dr. K. A. Jackson, now of the Bell Telephone Laboratories, formerly of the Division of Engineering and Applied Physics at Harvard; Mr. J. L. Walker, of the General Electric Research Laboratories; Professor David Turnbull, of the Division of Engineering and Applied Physics, Harvard University, previously of General Electric; and Professor M. C. Flemings of Massachusetts Institute of Technology, who have contributed so much to my understanding of various aspects of the subject. I would also like to acknowledge the financial support provided by the United States Atomic Energy Commission for some of the work reviewed herein. I am also indebted to the following for permission to use copyrighted material:

Pergamon Press
Deutsche Gesellschaft for Metallkunde
American Chemical Society
American Society for Metals
The Institute of Metals (London)
The University of Chicago Press
The Metallurgical Society of AIME
National Research Council of Canada
The Iron and Steel Institute (London)
American Institute of Physics
John Wiley and Sons
McGraw-Hill Book Company
Acta Metallurgica

B. C.

Contents

Principles
of Solidification

1

Introduction

1.0 General

Our direct knowledge of the nature and behavior of matter is almost always derived from the observation of samples that are very large compared to the size of a single atom; the measurements that we make, and the terms that we use, relate to the average behavior or properties of a very large number of atoms, whose individual behavior may, however, depart drastically from the average. While many of the properties of materials are controlled by the statistical average, there are some very important aspects of their behavior that depend upon the exceptions. The most convenient approach to the understanding of solidification phenomena is to consider them in terms of the macroscopic properties of materials, such as temperature, latent heat, composition, and surface free energy, and to consider separately how these properties, which are measures of the average or statistical behavior of very large numbers of atoms or molecules, can be accounted for in terms of the behavior of the individual atoms.

It is first necessary, therefore, to define the macroscopic properties that must be used, and to consider the relationships between them; this is an aspect of thermodynamics. It is not intended that this chapter should form more than a brief introduction to or reminder of the relevant thermodynamic laws, properties, and relationships.

1.1 Equilibrium between Solid and Liquid

If we define solidification as a process by which a solid grows at the expense of a liquid with which it is in contact, a useful point of departure is a consideration of the conditions under which a solid and a liquid can co-exist. If they can co-exist without any change in their relative quantities, they are said to be in equilibrium. This is a condition in which solidification does not occur; but experiment and theory both show that solidification does occur when the conditions depart only slightly from those of equilibrium.

1.2   Melting Point

Whenever a pure element or compound can exist both as a crystalline solid and as a liquid, there is a temperature, T_E, above which the liquid is the stable form of the material, and below which, the solid is stable. The temperature so defined is the *melting point* of the material. It is also the only temperature at which the crystalline solid and the liquid can co-exist in equilibrium. Table 1.1 gives the melting points of the common metals.

The melting point of a pure element or compound is, for most purposes, a constant, but it does vary slightly with pressure. This is because the application of a pressure tends to favor the formation of the phase (solid or liquid) which has the smaller specific volume. Most metals expand on melting, the solid being the denser phase. Increase of pressure, therefore, in such cases, raises the melting point. On the other hand, some substances, including water, gallium, germanium, silicon, and bismuth, contract on melting. Pressure lowers the melting point of such materials. The change of melting point corresponding to a change in pressure of one atmosphere can be calculated from the Clapeyron equation:

$$\frac{\Delta T}{\Delta P} = \frac{T_E(V_2 - V_1)}{L}$$

in which ΔT is the change of melting point in centigrade degrees resulting from a change of pressure ΔP in dynes/cm^2; T_E is the melting point (absolute); V_1 and V_2 are the volumes of 1 gm of solid and liquid respectively; and L is the latent heat of fusion in ergs/gm.

An example of the effect of pressure on the melting point is that of nickel, whose density changes, on melting, from 8.9 gm/cm^3 to 8.4 gm/cm^3. Using the accepted figures for *melting point*, 1728°K, and *latent heat of fusion*, 74 cal/gm, it follows that the melting point increases by 2.7×10^{-3} degrees for an increase in pressure of one atmosphere. This result is representative of those that would be found for other metals, in which the change in specific volume is generally comparable, and the value of T_E/L is, for reasons that are discussed in Chapter 2, nearly constant. The change is, therefore, rather small, so that, for example, the melting point of a metal in a vacuum is very close indeed to the value at atmospheric pressure, and the effect of ordinary variations of atmospheric pressure can be ignored for almost

Table 1.1. Melting Points of Some Elements

Metal	Melting Point		
	Fahrenheit	Centigrade	Kelvin
Aluminum	1220	660	933
Beryllium	2340	1280	1553
Chromium	3430	1890	2163
Cobalt	2723	1495	1768
Columbium	4380	2415	2688
Copper	1981	1083	1356
Gallium	86	30	303
Germanium	1760	958	1231
Gold	1945	1063	1336
Indium	314	156	429
Iridium	4449	2454	2727
Iron	2802	1539	1812
Lead	621	327	600
Lithium	367	186	459
Magnesium	1202	650	923
Mercury	−38	−39	234
Molybdenum	4760	2625	2898
Nickel	2651	1455	1728
Osmium	4900	2700	2973
Platinum	3224	1774	2047
Plutonium	1184	640	913
Radium	1300	700	973
Rhodium	3570	1966	2239
Silicon	2605	1430	1703
Silver	1761	961	1234
Sodium	208	98	371
Tantalum	5425	2996	3269
Tin	449	232	505
Titanium	3300	1820	2093
Tungsten	6170	3410	3683
Uranium	2065	1130	1403
Vanadium	3150	1735	2008
Zinc	787	419	692
Zirconium	3200	1750	2023

all purposes. Except for one special consideration (see Chapter 3), pressure effects will be ignored, and the melting point will be regarded as a constant for any given material.

1.3 Equilibrium of Alloys

Many applications of solidification processes take place in alloys rather than in pure metals, and it is therefore necessary to consider the melting point, or its equivalent, for an alloy. Attention will first be directed to single-phase alloys; that is, alloys in which equilibrium can exist between the liquid and a single type of crystal containing both (or all) the elements that are present. It is found experimentally that a single-phase alloy and the liquid in equilibrium with it are almost always of different compositions, and that the temperature of equilibrium depends upon the compositions. For each composition of the liquid there is a temperature, the *liquidus temperature,* at which it is in equilibrium with the appropriate solid, and, conversely, each composition of solid has a *solidus temperature* at which it is in equilibrium with the appropriate liquid. A solid and a liquid can be in equilibrium with each other only when they are at the same temperature, which must be the liquidus temperature of one and the solidus temperature of the other.

Equilibrium diagrams. Taking the alloys of copper and nickel as an example, the liquidus temperature is plotted as a function of composition in Fig. 1.1a and the solidus temperature is plotted in Fig. 1.1b. These two curves define the combinations of composition and temperature above which the liquid is stable, and below which the solid is stable, respectively. The two curves are combined in Fig. 1.2, which shows, by means of horizontal lines such as AB, which composition of the solid (S) and of the liquid (L) are in equilibrium at the same temperature (T), and, therefore, with each other. It follows that the solid can be in equilibrium with the liquid over a range of temperatures, but that for any given composition of either the solid or the liquid, the composition of the other phase and the equilibrium temperature are fixed.

It will be observed that there are combinations of composition and temperature that lie between the liquidus and the solidus curves. This is a reflection of the fact that the kinds of materials to which these considerations apply, such as metals, cannot exist in equilibrium in any state *between* that of the liquid and that of the solid; that is, there is no stable arrangement of the atoms that is intermediate between that of the solid and that of the liquid. If a crystalline

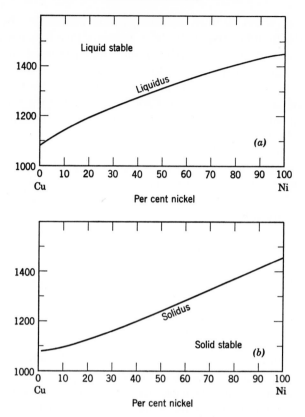

Fig. 1.1. Liquidus and solidus lines of the copper nickel alloy system. (a) Liquidus, (b) solidus. (From Ref. 2. Used by permission.)

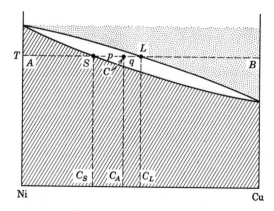

Fig. 1.2. Nickel-copper equilibrium diagram.

material has a temperature and composition represented by a point between the liquidus and the solidus curves, either it is not in stable equilibrium or it consists partly of solid and partly of liquid, each with the composition appropriate to the temperature. Thus, any point, such as C, in the region of the diagram between solidus and liquidus, corresponds to equilibrium between solid and liquid of the compositions C_S and C_L; the position of C determines the relative quantities of the solid and the liquid.

The composition C_A is the average composition of the alloy as a whole; it is, therefore, the average composition of a quantity L of liquid of composition C_L and a quantity S of solid of composition C_S. The ratio of these quantities can be determined as follows: the composition C_A is given by

$$\frac{S \times C_S + L \times C_L}{S + L} = C_A$$

from which

$$\frac{L}{S} = \frac{C_A - C_S}{C_L - C_A} = \frac{p}{q}$$

This is known as the *lever rule*.

Noncrystalline materials such as glasses and some polymers undergo a smooth, continuous transition between the liquid and solid form. The considerations in this book are limited to those substances in which there is an abrupt discontinuous change in structure and properties on passing from solid to liquid. Relatively few alloy systems are as simple as the one represented in Fig. 1.2, although the main feature, equilibrium between a solid of one composition and a liquid of another at a fixed temperature, is common to all alloy systems, whether they contain only two elements, as in the case discussed above, or more. Many systems, however, also show one or more of the following special equilibrium relationships.

Congruent melting. There are a relatively small number of binary alloy systems in which the solid and the liquid in equilibrium with it have the same composition; this can occur only at a fixed composition and temperature. Examples of two types are shown in Figs. 1.3a and 1.3b. The point C, where equilibrium exists at identical compositions, is described as the point of congruent melting.

Eutectics. A eutectic point occurs at the intersection of two liquidus lines (L_1 and L_2 of Fig. 1.4a) that slope in opposite directions; at this point A, which is at a fixed temperature and composition, the liquid is in equilibrium with two different solid phases, which may be

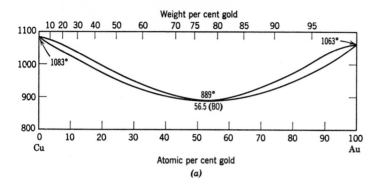

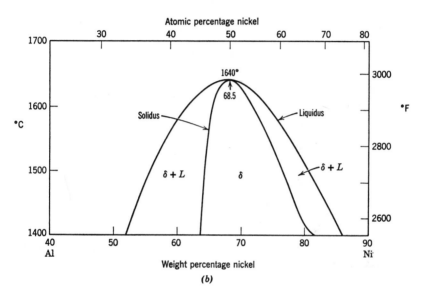

Fig. 1.3. Congruently melting alloys. (a) Minimum type, (b) maximum type. [Part (a) from Ref. 2, p. 199; part (b) from Ref. 4, p. 1164. Both used by permission.]

terminal solid solutions, as in Fig. 1.4a, congruently melting phases, or a terminal solid solution and a congruently melting phase as in Fig. 1.4b.

Peritectics. A peritectic point P exists at the intersection of two liquidus lines (L_1 and L_2 of Fig. 1.5) that slope in the same direction. Here again, a liquid is in equilibrium at a single temperature with two solid phases.

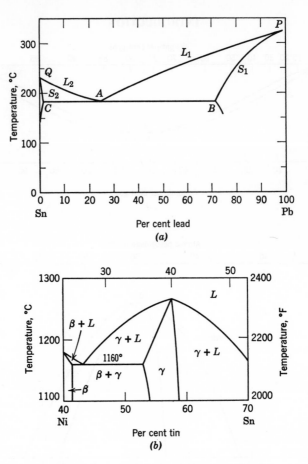

Fig. 1.4. Eutectic equilibrium. (From Ref. 4, p. 1234. Used by permission.)

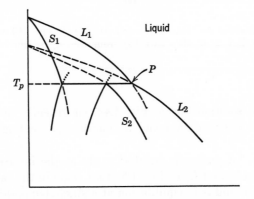

Fig. 1.5. Peritectic equilibrium.

8

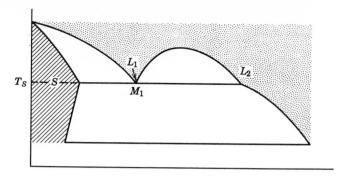

Fig. 1.6. Monotectic equilibrium.

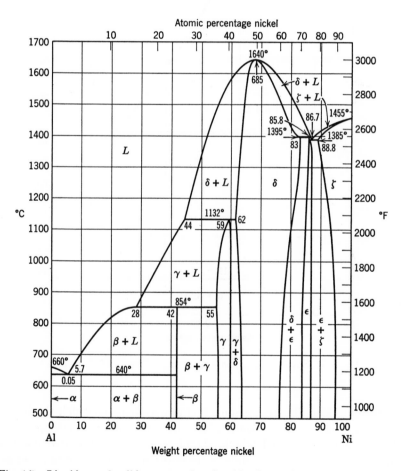

Fig. 1.7. Liquidus and solidus curves for the Aluminum-nickel system. (From Ref. 4, p. 1164. Used by permission.)

9

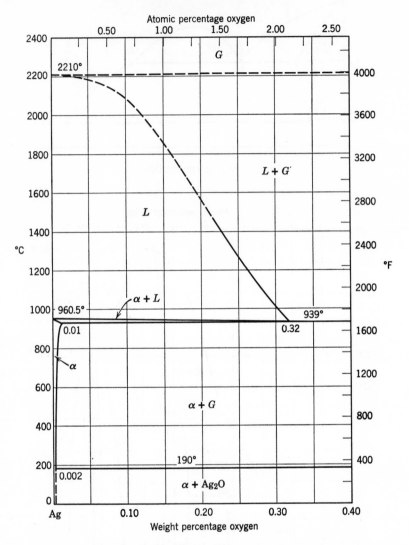

Fig. 1.8. Silver-oxygen phase diagram. (From Ref. 4, p. 1152.)

Monotectics. There are some metallic systems in which the liquid is unstable as a single liquid phase at temperatures that approach the freezing range, and which, therefore, split up into two distinct liquid phases. Because of differences in density, these tend to separate into two layers. The temperature at which one of the liquids begins to solidify is the *monotectic temperature;* the typical monotectic diagram is shown in Fig. 1.6. At the monotectic point M_1 the liquid L_1 is in equilibrium with the liquid L_2 and the solid phase S.

The complete phase diagram for a binary alloy system may contain any or all of the features described in the preceding paragraphs; an example is shown in Fig. 1.7.

1.4 Gas–Metal Equilibrium

A special, but sometimes important, case of binary equilibrium in metals is that between a gas, usually oxygen, hydrogen or nitrogen, and the metal. The equilibrium of a metal with its own vapor will not be considered here. There are two main types of gas–metal equilibrium: (a) those in which the liquid and the solid each take the form of solid solutions, and (b) those in which a compound phase is formed. An example of the former type is shown in Fig. 1.8, and the latter in Fig. 1.9. In each case the phase diagram is shown for a fixed pressure of the gas.

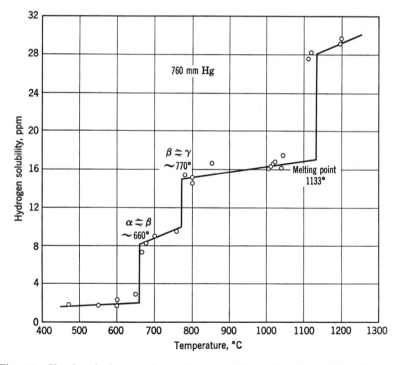

Fig. 1.9. Uranium-hydrogen phase diagram. (From Ref. 2, p. 803. Used by permission.)

1.5   Ternary and Multicomponent Alloys

The considerations outlined above apply to alloys containing not more than two elements; they also apply in part, however, to alloys with more than two components. In particular, any alloy has a unique liquidus temperature for each possible composition of the liquid, and similarly, it has a unique solidus temperature for each composition of the solid. An important difference from the two-component case is that, in systems of three or more components, the solidus and liquidus compositions are not uniquely defined, in general, by the temperature. The simplest example is that of the ternary system; compositions in ternary systems are usually represented as points on or in an equilateral triangle, *ABC* of Fig. 1.10). The corners of the triangle correspond to the three pure components, and the edges represent the three binary systems. Each point within the triangle corresponds to a definite composition; the ratio of the three components for the point *P* is equal to the ratio of the distances *a*, *b*, *c* of *P* from the three sides of the triangle. The sum of these three distances is constant for a given equilateral triangle. Since each possible composition of the liquid corresponds to a point on or in the triangle, it follows that the liquidus is a surface, rather than a line as in the "two-component" case. In a three-component system therefore there must be a *line* on a liquidus surface that corresponds to any particular temperature,

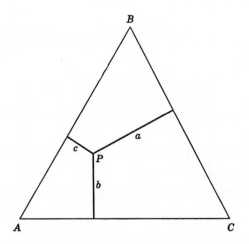

Fig. 1.10. Representation of ternary composition.

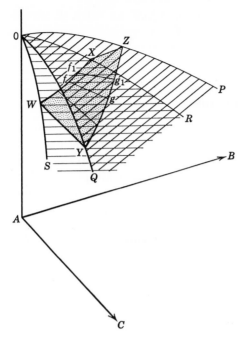

Fig. 1.11. Liquidus surface, solidus surface, and tie-lines of a ternary system.

and therefore a knowledge of the liquidus surface and the solidus surface does not define which solidus composition is in equilibrium with a specified point (i.e., composition) on the liquidus. It is therefore necessary to use *tie lines* to indicate the equilibrium relationships. An example is given in Fig. 1.11 in which one corner of a ternary diagram is shown. The two binary liquidus lines, OP and OQ are joined by the liquidus surface, and similarly, the binary solidus lines OR and OS are joined by the solidus surface. The lines WX and YZ indicate the points on the two surfaces at which the temperature has a fixed value T_1. Some points on WX and their equilibrium points on YZ are joined by the tie lines fg, f_1g_1, etc. Similarly, any given composition of the liquid in a quaternary or higher alloy has a unique liquidus temperature which may be shared by a continuous range of compositions.

The composition of a four-component alloy could, in principle, be represented by a point in a tetrahedron, by analogy with the representation of a ternary composition by a point in a triangle; but there is no possible geometrical method for representing quaternary or higher phase diagrams, in which an extra dimension is required to represent temperature. For a more detailed discussion and further examples of phase diagrams, the reader is referred to more specialized books (1–4).

1.6 The Phase Rule

An equilibrium diagram shows in graphical form the nature of the phase, or phases, that may be present when an alloy of any given composition is in equilibrium at a given temperature. It therefore represents the phase or phases that are present when the system has the lowest possible free energy; that is, the condition in which any change of phase is necessarily accompanied by an increase in free energy. The equilibrium between alloy phases is subject to various restrictions, the recognition of which is an important aspect of thermodynamics. The most basic of these restrictions is expressed as the *phase rule*, which may be written as

$$P + F = C + 2$$

when C is the number of components in the system; that is, the number of chemical substances which can be varied separately. F is the number of degrees of freedom, that is, the number of parameters such as pressure, temperature, and composition that can be varied independently of each other, and P is the maximum number of phases that are present in the system at equilibrium.

If, for example, there is only one component, $C = 1$ (a pure metal), then there can be three phases ($P = 3$) only when $F = 0$, i.e., none of the parameters can be varied; thus there is only one temperature and pressure at which the solid, liquid, and vapor forms of a pure substance can be in equilibrium with each other. This is called the *triple point.* Similarly, for a binary alloy ($C = 2$) there are three degrees of freedom for the existence of a single phase; that is, the solid can exist over a range of pressures, temperatures, and compositions; however, solid and liquid can exist together ($P = 2$) only under more restricted conditions; if the pressure and temperature are specified, then the composition must have a fixed value; there are only two degrees of freedom. Again, at the eutectic point, the liquid, two solid phases, and the vapor would all be in equilibrium; then $P = 4$, $C = 2$, and $F = 0$; in other words, there is only one combination of composition, temperature, and pressure that can give four-phase equilibrium. The application of the phase rule to ternary and higher systems is similar; a ternary eutectic, for example, has no degrees of freedom, but equilibrium between a solid phase, the liquid and the vapor ($P = 3$), in a ternary system has two degrees of freedom; for a fixed pressure, both temperature and composition can vary independently; this does not mean that these two conditions can have *any* value, but it does mean that when a value of, say, temperature, has been selected, there is a range of values of composition that satisfy the equilibrium conditions. For many, but not all, purposes relating to solidification, the pressure may be regarded as constant; when this is valid, it is convenient to use the form $P + F = C + 1$ (constant pressure), which restricts the variables to be considered to temperature and composition. It can easily be seen that this form of the phase rule gives the same result as the more general form.

1.7 The Distribution Coefficient

For many purposes, it will be sufficient to consider two-component systems; in such systems, the liquidus and solidus lines define the equilibrium relationships between solid and liquid. Many of the most significant aspects of solidification at the "microscopic" level are related to the fact that the composition of the solid that is formed is different from that of the liquid immediately in contact with it, and therefore at or close to equilibrium with it. A convenient way of describing this difference is by means of the ratio of concentration of solute in the solid to that in the liquid in equilibrium with it. This is the ratio AB/AC in Fig. 1.12. This is referred to as the *equilibrium distribution coefficient* k_0. While k_0 is usually not independent of composition in a given alloy, it is often sufficiently constant over a wide range to be treated as if it were constant.

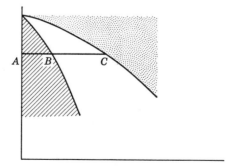

Fig. 1.12. The equilibrium distribution coefficient.

Most of the binary phase diagrams for metals have been determined, and are readily available (1, 2, 3); however, some of these diagrams are in error, partly owing to the use of materials of low purity for their determination, and partly because of the inherent difficulty of determining the solidus curve. A useful method of checking the accuracy of the slope dC_S/dT_S of the solidus line from that of the liquidus (which is more reliable) is by means of the van't Hoff equation, which is a thermodynamic expression for the difference of slope of the solidus and liquidus lines in terms of the melting point T_E of the solvent and its latent heat of fusion only:

$$\frac{dC_L}{dT_L} - \frac{dC_S}{dT_S} = \frac{L}{RT_E{}^2}$$

This equation applies strictly only at very low concentrations; that is to say, it relates to the slope of the liquidus and solidus lines close to the point where they meet on the "pure metal" axis. It is of interest to note that the solute metal does not enter into this expression; in other words, the difference between the solidus and liquidus slopes, in centigrade degrees per unit change in atomic concentration, is constant for a given solvent metal. In the case of copper, for example,

Atomic weight = 63.5

L = 50.6 cal/gm or 3280 cal/mole

Melting point = 1083°C = 1356°K

R = 1.98 cal/mole °K

Hence

$$\frac{dC_L}{dT_L} - \frac{dC_S}{dT_S} = 0.0009$$

The accepted phase diagram for the copper-zinc system (Fig. 1.13) has an initial liquidus slope of about 28 atomic per cent per 100 degrees and a solidus slope of about 20 atomic per cent per 100 degrees. These become

$$\frac{dC_L}{dT_L} = 0.0028 \quad \text{and} \quad \frac{dC_S}{dT_S} = 0.0020$$

from which the difference is 0.0008, in fair agreement with the predicted value of 0.0009; on the other hand, the copper zirconium diagram (4) (Fig. 1.14) has initial liquidus and solidus slopes of

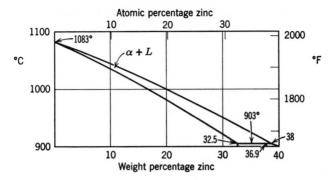

Fig. 1.13. Solidus and liquidus lines for solid solutions of zinc in copper. (From Ref. 4, p. 1206.)

0.00133 and 0.00009 atomic per cent per degree, giving a difference of 0.00124, which disagrees substantially from the value of 0.0009 for any copper alloy given by the Van T'Hoff equation. It is probable that the solidus curve is inaccurate.

1.8   Thermodynamic Criteria for Equilibrium

For some purposes it will be necessary to express the conditions for equilibrium in formal thermodynamic terms. This is based on the definition of a property of a system that can spontaneously decrease, but cannot increase unless external work is supplied. This property is the *free energy*, which is the sum of the free energies of all the parts into which the system can be subdivided. For reasons that are

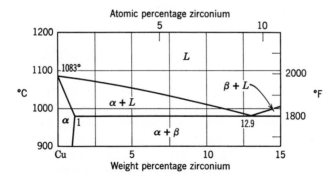

Fig. 1.14. Solidus and liquidus lines for solid solutions of zirconium in copper. (From Ref. 4, p. 1207.)

outside the scope of this book (5) the Gibbs free energy G is defined as

$$G = H - TS$$

where H is the enthalpy, T the absolute temperature, and S the entropy. Enthalpy is equal to the internal energy E plus the product of the pressure and the volume, or $H = E + PV$. If the effect of pressure can be ignored, it is often more convenient to use the Helmholtz free energy $F = E - TS$.

If a pure material can exist in either of two phases, the stable one is that which has the lower free energy, because the free energy of the system would be increased if any of the material changed to the other phase. It follows that the condition for equilibrium between two pure phases is that

$$E_L - T_E S_L = E_S - T_E S_S$$

where the subscripts L and S refer to liquid and solid, respectively and T_E is the temperature of equilibrium (i.e., the melting point). The equilibrium temperature, therefore is that at which $E_L - E_S = T$ $(S_L - S_S)$, or $\Delta E = T_E \Delta S$; $E_L - E_S$ is the change of internal energy on melting, and is therefore the latent heat of fusion. The entropy of melting ΔS is equal to $\Delta E / T_E$.

The free energies of the two phases (solid and liquid) are equal only at one temperature, the melting point. (This must be so unless ΔE is zero, in which case ΔS is also zero and there is no thermodynamic difference between the phases.) At other temperatures, the difference of free energy per unit mass is calculated as follows:

$$F = E - TS, \quad \text{or} \quad \Delta F = \Delta E - T\Delta S$$

This is equal to zero when $T = T_E$, the melting point. Therefore $\Delta S = \Delta E / T_E$, and, if it is assumed that ΔE (the latent heat of fusion) and ΔS (the entropy of fusion) do not change with temperature, then

$$\Delta F = \Delta E - T \frac{\Delta E}{T_E} = \Delta E \frac{(T_E - T)}{T_E} = \frac{\Delta E \Delta T}{T_E}$$

where ΔT is the departure from the temperature of equilibrium T_E.

In the special case of a pure substance, considered above, the criteria for equilibrium, that of minimum free energy for the system as a whole, also requires that the free energies of the two phases (per unit mass) be equal. This is not true in the more general case in which the compositions of the two phases may be different from each other. The criterion for equilibrium for a multicomponent system is that the chemical potential is the same in both phases for each component. The reader is referred to more specialized books for a detailed discussion of equilibrium in binary or more complex alloys.

References

1. F. N. Rhines, *Phase Diagrams in Physical Metallurgy*, McGraw-Hill Book Company, 1956.
2. M. Hansen, *Constitution of Binary Alloys*, McGraw-Hill Book Company, 2d Ed., 1958.
3. C. Smithells, *Handbook of Metals*, Interscience, New York, 1955.
4. *Metals Handbook*, American Society for Metals, Cleveland, 1948.
5. R. Swalin, *Thermodynamics of Solids*, Wiley, 1962.

2

Solidification
as an Atomic Process

2.1 Solids and Liquids

While many of the features of solidification can be understood in terms of bulk properties such as thermal conductivity, in which the contributions of individual atoms are ignored, a deeper understanding requires their description in terms of the properties and behavior of the atoms themselves. The explanation of bulk properties and behavior of matter in terms of the behavior of the single atoms is the province of Statistical Mechanics, a subject in which refined physical models are examined by means of sophisticated mathematical methods that are far beyond the scope of this book. It will be sufficient for our purposes to use the simplest possible treatment, with a view to developing a general, rather than a precise, understanding of solidification. A precise understanding of a number of aspects is not yet within sight.

The nature of crystalline solids. For our purposes it is sufficient to regard a crystal as a regular assembly of atoms, which may or may not be all of the same kind. Each atom has a site, defined by its geometrical relationship with the crystal lattice, which is the mean position of the center of the atom. The actual position of the atom is less well defined, because it undergoes thermal vibration within the space available to it between its neighbors. Its mean position is one of minimum energy, that in which the electrostatic (and sometimes magnetic) forces exerted on it by the other atoms just balance.

The thermal energy of the crystal may be regarded as the sum of the individual vibrations of all the atoms, each atom exchanging kinetic energy of motion for potential energy of attraction and repulsion, as it moves away from some of its neighbors and toward others. Since all the atoms are doing this, the energy of an individual atom is likely to change each time it collides with a neighbor. The average energy of the atoms is equal to $\frac{3}{2}kT$, where T is the absolute temperature and k is Boltzmann's constant, 1.38×10^{-16} ergs/°K. The average frequency of their vibration is represented by ν.

Alternatively, the thermal motion of the atoms in a crystal can be thought of as the result of a large number of independent waves propagating through the crystal. These waves, which are longitudinal elastic pressure waves, are analogous to sound waves, and are called *phonons*. This way of describing the thermal motion of the atoms emphasizes the fact that the energy of the thermal motion is passed on from one atom to the next (this is the essence of wave motion) and that each atom is not in its own independent state of vibration. Whichever way is preferred for describing the motion of the atoms, it follows that there is a wide range of individual energies at any instant, and that the energy of an individual atom is likely to change each time its direction of motion is reversed.

At any instant, the energy is distributed among the atoms or molecules according to the Boltzmann Distribution, which is represented analytically by the relationship

$$\frac{n_1}{n_2} = \frac{\exp\ (-E_1/kT)}{\exp\ (-E_2/kT)}$$

which gives the equilibrium ratio of the numbers n_1 and n_2 of atoms (assumed to be identical) with energies E_1 and E_2, at the temperature T, and is illustrated graphically in Fig. 2.1, which shows how the number N_E of atoms which have energies within a given range varies with the mean energy E in that range.

The mean energy E_M of all the atoms is equal to $\frac{3}{2}kT$, and therefore defines the temperature of the crystal. An important characteristic

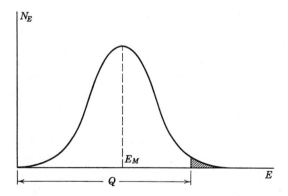

Fig. 2.1. Boltzmann distribution of energies.

of the distribution is that at the temperature T the proportion of atoms that have energies above some value Q is equal to exp $(-Q/kT)$. Thus if Q is the energy that an individual atom must have in order to leave its "potential well," and move its mean position to a new site, then the proportion of atoms having this energy is known, for any given temperature. If the temperature is high, so that kT is comparable with Q, then the proportion is high; it decreases rapidly as the temperature decreases. Further, if each atom vibrates ν times per second, and if the energy of each vibration is independent of that of the previous one, then the number of times per second that any individual atom is likely to have energy equal to or greater than Q is equal to $\nu \exp(-Q/kT)$.

A complete description of the crystalline state would also include the imperfections that are known to exist in crystals; in addition to the external surface, which may be regarded as an imperfection because the ideal crystal would be infinite in extent, there are vacancies, interstitials, dislocations, stacking faults and sub-boundaries, of which vacancies must, and the others may, be present in a crystal.

Nature of liquids. It is much more difficult to describe liquids from the atomic point of view, and there are marked divergences of opinion as to the most appropriate way to do so; one approach is to regard the liquid as a random array of atoms all undergoing thermal vibration, or transmitting phonons. The configuration, when averaged over times that are long compared with $1/\nu$, the time of one vibration, appears to be random; this is the kind of information that can be derived from x-ray diffraction studies, for example. Randomness, however, needs closer description, because the simplest type of random structure, in which the centers of the atoms are distributed in a purely random way, would require some atoms to be closer to each other than is believed to be possible. A second type of random structure is that in which the randomness is restricted, in the sense that no two centers may be closer together than the atomic diameter, combined with the condition that there are nowhere spaces large enough to insert extra atoms. This specification would give the observed x-ray patterns, at least qualitatively, and would account in general terms for many of the observed properties of liquids, such as the viscous behavior, electrical, thermal, and diffusion properties, and density. While it is probably true that the time average of the atomic positions is random, within the restrictions described above, it may also be true that there is, at any instant, substantial short range order, in the sense that each atom may be related to some, or even many, of its neighbors exactly as it would be in the crystal. The x-ray diffraction data are compatible with the interpretation that the interatomic spacing is very similar to that of the crystal, but the number of nearest neighbors is rather smaller. Thus the instantaneous structure might be one in which each atom is part of a crystal-like "cluster," oriented randomly, and with some "free space" between it and its neighboring clusters, with some of which it might share an atom. These clusters would form and disperse very quickly, by transfer of atoms from one to another by movement across the intervening free space. It is instructive to consider these clusters from a thermodynamic point of view; the liquid is regarded as an equilibrium mixture of "molecules." The term "molecule" is used here to represent a single atom or any possible grouping of two or more atoms. The relative numbers of the different "molecules" is given by

$$\frac{n_i}{n} = \exp\left(\frac{-\Delta G}{kT}\right)$$

where n_i is the number of "molecules" containing i atoms, n is the total number of atoms, and ΔG is the excess free energy of the "i molecule" compared with the same number of single atoms. It is evident from the formula that as ΔG increases, the number of molecules decreases.

The difference of free energy ΔG arises from two sources; one depends on ΔG_V, the difference in free energy between solid and liquid; the other, ΔG_S is the result of the free energy of the surface separating the solid from the liquid. The former component, ΔG_V, is zero at the equilibrium temperature, negative below it, and positive above, while ΔG_S is always positive. The value of ΔG, therefore is given by

$$\Delta G = V \Delta G_V + A \Delta G_S$$

where V is the volume of the "molecule," and A the area of its surface. If the cluster is large enough to be regarded as a sphere, then

$$\Delta G = \frac{4}{3} \pi r^3 \Delta G_V + 4\pi r^2 \Delta G_S$$

The existing evidence (which is discussed more fully on page 75) shows that, in liquid metals near the melting point, the values of the quantities in this equation are such that "molecules" containing more than a few hundred atoms should never occur, in a sample of, say one cubic centimeter volume. In a much larger sample of 10^{15} times the size (1 cubic kilometer) there would be a good probability of a substantially larger cluster existing. It has been assumed so far that the cluster is "solid-like," in the sense that its atoms are related to each other as they would be in a crystal of the same material as the liquid. The reason for the selection by a material of its characteristic crystal structure is that this is the structure that has the lowest free energy; but this implies that the same atoms could be arranged in any other structure, at the cost of some free energy. The foregoing discussion shows that clusters will exist even when their free energy is higher than that of the liquid. It must be concluded that clusters of all possible crystal structures should exist, but that the structures that have higher free energies will be represented by fewer clusters, of a given size, than the more favorable structures.

A somewhat different point of view is that of Frank (1) and Bernal (2), who suggest that it is necessary to invoke special nonrandom components of structure that are not repeatable to fill space, whereas the essence of a crystal structure is that unlimited repetition is possible. According to the thermodynamic argument, these special structures should exist, but only in the proportion dictated by their free energy.

While there is some doubt as to the best way to describe the structure of liquids, it is clearly valid to consider a liquid as an assemblage of atoms or molecules, vibrating with mean energy $\frac{3}{2}kT$ and an average frequency ν; however, while the neighbors of any individual atom in a crystal remain unchanged for relatively long periods, the neighbors in a liquid change much more frequently. This is because there is much more "free space" in a liquid than in a crystal, and this means that an atom in a liquid is *not* contained in a "box" of other atoms from which it can escape only by filling a vacancy that happens to arrive at one of the sites that form the box. In the liquid, the box is less well defined and an atom can move its mean position by a much smaller distance than is necessary in the crystal. The evidence for this statement comes from the viscous behavior of liquids, and their diffusion characteristics.

The difference between crystals and liquids. The most obvious difference between crystals and liquids is their behavior when subjected to stress; a liquid changes shape in response to any shear stress however small, while a solid is deformed elastically by a shear stress, in the sense that some or all of the shear strain disappears as the stress is removed. The difference, which is often used to define the difference between solids and liquids, is of little importance in connection with solidification; much more significant is the fact, known for nearly two hundred years, that heat must be added to a crystalline solid at its melting point to convert it into liquid at the same temperature; this is the *latent heat of fusion*. It should be pointed out that the latent heat that must be added to a crystalline solid to convert it into liquid does not increase the energy of thermal vibration, which is still $\frac{3}{2}kT$ per atom or molecule; it is used to relocate the atoms in positions which have, in general, higher potential energy, in the sense that while each atom in the liquid is located at a free energy minimum, these minima are higher than those of the solid.

A second difference that is important in the present context is the difference in density between a crystal and the liquid formed by melting it; this can be understood in terms of the difference of structure; when the crystal has a close-packed or nearly close-packed structure (such as face-centered cubic, close-packed hexagonal or body-centered cubic) then random (or more random) packing of the same number of atoms, which are themselves unchanged, requires more volume, and the density of the liquid is lower than that of the solid. This applies to all the metals except gallium and bismuth, which are "borderline" cases for classification as metals. On the other hand, when the crystal has a less close-packed (or more "open") structure, the density of the liquid is higher than that of the crystal. This is the case for germanium and silicon as well as gallium and bismuth.

Quasi-chemical approach. A useful, but not rigorous, way of looking at the difference of energy of the solid and liquid is by the *quasi-chemical approach,* which depends on the basic assumption that the whole crystal (or liquid) is held together by forces between pairs of atoms; a consequence is that the energy required to vaporize a crystal is the sum of the energies required to break all these "pairwise" bonds. We will assume for our present purpose that the only pairwise interactions that need to be considered are those between nearest neighbor atoms. For convenience we will restrict the discussion to the case of face-centered cubic crystals, although the conclusions are of much more general application.

Taking copper as a specific example, it is found experimentally that the heat of vaporization is about 80 Kcal/mole, while the heat of fusion is about 3.1 Kcal/mole. This indicates that it takes about 25 times as much energy to separate the atoms completely from each other as it does to change the structure from solid to liquid. In the face-centered cubic structure, each atom has 12 nearest neighbors, and therefore 12 "nearest neighbor bonds," each of which is shared by one other atom. There are, therefore, 6 times as many bonds as there are atoms, if for the moment we disregard the atoms at the surface, which have fewer nearest neighbors. The latent heat per atom is therefore equal to the work that must be done in breaking 6 bonds, that is, to the energy of 6 bonds. In the case of vaporization, discussed so far, the concept is unambiguous; but in the case of melting, the bonds are not destroyed completely; in fact only about $\frac{1}{25}$ of that energy is required. It would be reasonable to assume that, during melting, either 1 bond in 25 is broken, or that each bond is replaced by one with about 4 per cent less energy. We will adopt the latter assumption, realizing that it represents *average* behavior because the nearest neighbor distances in the liquid are not equal and the bond energies, therefore, are also not equal.

It will be assumed that the energy required per atom to change from solid to liquid is proportional to the number of *"crystal bonds"* that are broken, that is, the decrease in number of nearest neighbors *in the crystal*. The heat of fusion per atom corresponds to a decrease of 6 in the number of crystal bonds. These are replaced by the weaker *liquid bonds*. Thus, the heat required to melt a crystal is equivalent to 6 "crystal bonds" per atom.

2.2   The Solid-Liquid Interface

The atoms at the surface must now be considered. It is assumed that the crystal is undistorted right up to its surface, and that all the bonds have the same energy as the bonds within the crystal. An atom at the surface may have any number of nearest neighbors from 3 to 11, depending on how many of the neighboring sites are occupied. This is illustrated in Fig. 2.2.

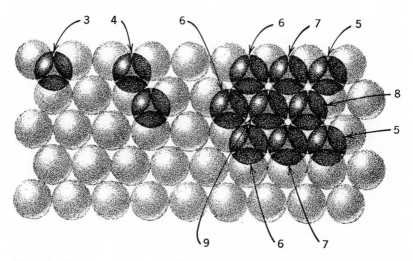

Fig. 2.2.   Number of nearest neighbors N of atoms on close-packed surfaces.

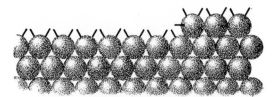

Fig. 2.3. Number of "bonds" per surface atom.

The energies of these atoms are higher than that of atoms in the interior, in proportion to the number of missing bonds; thus an atom with 6 neighbors will have an energy that is greater by half of the latent heat of fusion than if it were in the interior of the crystal. In order to remove it from the surface, it must be given the other half (its 6 remaining bonds must be broken). This reduces by 1 the number of nearest neighbors of each of 6 adjacent atoms, each of which therefore is given $\frac{1}{12}$ of its latent heat; i.e., a total of one-half of the latent heat is shared between 6 atoms that remain. Thus the removal of 1 atom from a site of this kind requires exactly the latent heat per atom, and the "missing bond" energy of the atoms forming the surface (i.e., the surface energy) has not changed. The removal or addition of an atom with 6 crystal neighbors is the only case, for the face-centered cubic structure, in which the surface energy is not changed, and it is the only case in which exactly the latent heat is required; it is called the "repeatable step." Consider another example, that of an atom with 4 crystal neighbors. It has $\frac{8}{12}$ of the latent heat, and on removal requires the remaining $\frac{1}{3}$. When it is removed, 4 atoms each have their energy raised by $\frac{1}{12}$ of the latent heat; the total energy expended is $\frac{2}{3}$ of L (the latent heat per atom) and the energy of the surface has been reduced by $\frac{1}{3}L$.

The magnitude of the energy associated with the surface of a crystal in contact with its melt can now be estimated. If the surface consists of a smooth atomic plane of the close-packed type, each atom will have 9 crystal neighbors, and its energy will therefore be $\frac{1}{4}$ of the latent heat. If the surface is not smooth, but has "steps," as shown in Fig. 2.3, the average number of bonds per surface atom is not changed, because the bonds gained at exterior corners are lost at re-entrant ones; but the area over which these bonds are distributed is decreased slightly. For a cube face, {100}, in which the atoms are arranged as shown in Fig. 2.4, each atom in the surface plane has 8 crystal neighbors, and therefore the energy per atom is $\frac{1}{3}$ of the latent heat. However, there are fewer atoms per unit area in this

Fig. 2.4.   Positions of the atoms on a cube face of an F.C.C. crystal.

case than for the close-packed face, the ratio being 1:1.15. The energies per unit area are therefore in the ratio $1.15 \times \frac{1}{4} : 1 \times \frac{1}{3} = 1:0.87$. The energy per unit area is, therefore lower for the close-packed face than for the cube face, and it can be shown similarly that all other faces have higher energies.

The only experimental evidence for the energies of solid-melt interfaces is somewhat indirect; it is derived from nucleation experiments, discussed in Chapter 3. The values obtained in this way are somewhat higher than that calculated above, being about $0.4L$ per atom. The difference may be caused by the fact that the nucleation experiments are necessarily related to crystals that are so small that the external edges and corners, which have more "free bonds," contribute substantially to the surface energy of the crystal. Agreement between the experimental results and the simple nearest-neighbor theory outlined above is better than might be expected, perhaps because the contributions of more distant neighbors would be very similar in the crystal and in the liquid, and they can therefore be neglected.

Micro-topography of the interface. If the interface between a solid and a liquid is defined as the surface which separates all those atoms that occupy lattice sites from those that do not, then it is evident that the interface may be, but need not be, atomically smooth. The two possibilities are illustrated in Fig. 2.5, where (a) represents a smooth interface with a step, while (b) represents a rough interface; in each case the black circles represent the atoms of the liquid. It would be possible to describe the rough interface (b) as a smooth interface to which the liquid is locally related epitaxially; this, however, is excluded by the definition formulated above.

It has been shown theoretically (3) and by experiment that many crystals in contact with their vapors or with saturated solutions have smooth interfaces; the case of a liquid in contact with its own melt is different, mainly because the concentration of the appropriate atomic species in this case is approximately the same in the liquid as in the solid, whereas in solutions and vapors the species is much more dilute in the noncrystalline phase. A detailed discussion of these cases would not be appropriate here, but the basis of the argument is as follows: an atom of vapor arriving at a crystal surface may "stick" in the sense that it loses part of its heat of vaporization, and is then able to move about on the surface until one of three events occurs. It may acquire sufficient thermal energy to evaporate, it may find a site, on the edge of an existing step, at which the number of neighbors is sufficient to "trap" the atom by sufficiently lowering its energy; or it may encounter sufficient other adsorbed mobile atoms for them to stabilize each other by the formation of an "island" of a new layer. Theory shows that the last can occur only if the concentration of atoms on the surface is high. This requires high supersaturation in the case of a vapor, because otherwise the rate of arrival of atoms would be too low; it is also shown that under these conditions the surface tends to become smooth, although a surface newly formed by fracture, for example, would at first be rough. Growth of such a crystal would clearly be by lateral extension of existing islands and by building onto existing steps, each of which would decrease the roughness of the surface, because each step would grow until it reached the limit of the space available to it; it would either meet and coalesce with another step, or it would reach an intersecting

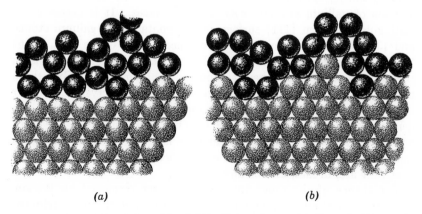

(a)   (b)

Fig. 2.5. "Smooth" and "rough" solid-liquid interfaces. (a) Smooth, (b) rough.

surface and cease to exist as a step. If all the steps are eliminated by growth, then further growth can take place only by the formation of new steps, which cannot occur without much higher supersaturation than is actually found to produce growth. This apparent anomaly was resolved by Frank (4), who showed that the presence of a dislocation with a component of its Burgers vector normal to the crystal surface provides, around its point of emergence, a perpetual step that can move but which cannot be eliminated by growth. The spiral step that was predicted for this mechanism has been observed in numerous crystals.

The evidence, both theoretical and experimental, suggests strongly that the interface between a crystal and its melt may be smooth or rough according to the characteristics of the material and the crystallographic nature of the surface. The experimental evidence is that some crystals grow from their own melt with surfaces that are atomically smooth and crystallographically significant; this group includes many organic materials, germanium and silicon under certain conditions, and ice on surfaces parallel to the basal plane of the structure but not on any other surfaces. On the other hand, metals normally solidify with surfaces that are not purely crystallographic, but are controlled at least in part by local thermal conditions. If the interface is smooth, and there is some difficulty in starting new layers, then the crystal should have faces that are flat and large in extent; if the interface is rough, then the growth will be much more sensitive to small differences of temperature. The detailed interpretation of this kind of growth is discussed in Chapter 4.

The influence of the atomic structure of the interface on its morphology during growth is strikingly displayed by the behavior of crystals of gallium arsenide with $\{111\}$ surfaces. It is necessary to distinguish between $\{111\}$ surfaces which contain only gallium atoms, and $\{\bar{1}\bar{1}\bar{1}\}$ surfaces, terminating in arsenic atoms. Gatos *et al.* (5, 6) have found that more perfect crystals are obtained when growth is on the "arsenic face" than when it is on the "gallium face." Growth on the gallium surface often gives rise to twins and stray crystals. Booker (7) has found that the growth steps on opposite faces of a ribbon-like dendrite of GaAs are quite different. On the gallium face, the steps are serrated, while those on the arsenic face are straight and much more regular. The serrated steps on the "gallium" face are so inclined that they are, in fact, arsenic faces.

The theory of the rough interface has been discussed by Burton, Cabrera, and Frank (8) and by Mullins (9), but the most relevant discussions of the nature of the interface of a crystal in contact with its own melt are those of Jackson (10, 11, 12), who points out that the equilibrium surface structure is that which is in equilibrium with both the crystal and the melt; Jackson assumes that atoms are added randomly to a surface that was initially atomically smooth, and calculates the change in surface free energy ΔF_s, as follows:

$$\Delta F_s = \Delta E_0 - \Delta E_1 + T\Delta S_0 - T\Delta S_1 - P\Delta V$$

where ΔE_0 is the decrease in energy corresponding to the addition of N_A single atoms to the interface from the liquid.

ΔE_1 is the average decrease of energy of the N_A atoms due to the presence of the other atoms on the surface (2 atoms in neighboring sites would have an energy lower by L/b than if they were separated).

ΔS_0 is the difference in entropy between the solid and the liquid.

ΔS_1 is the entropy that corresponds to the degree of randomness of the distribution of the N_A atoms on the surface, and $P\Delta V$ is the term arising from change in volume during the change in state; this term is negligible in the liquid-solid transformation.

The case considered by Jackson is that in which atoms can exist, as part of the crystal, in only 1 plane above the last complete plane. This is referred to as the "single-layer rough interface."

It is assumed, following the nearest neighbor approximation used above, that the energy change ΔE_0 is equal to $2L_0(\eta_0/v)N_A$, where L_0 is the latent heat per atom, η_0 is the number of nearest neighbors it actually has, while v is the number it would have in the interior of a crystal of the same structure. If there are N possible sites, then, on the average, each atom will find N_A/N of its nearest neighbor sites filled; if the number of nearest neighbor sites is η_1, then $E = (L_0\eta_1/v)(N_A{}^2/N)$, also $\Delta S_0 = (L/T_E)N_A$, since L/T_E is the entropy of melting per atom.

If N_A atoms are arranged at random among N sites, the probability W of any such arrangement is

$$W = \frac{N!}{N_A!(N - N_A)!}$$

By using Stirling's approximation and the Boltzmann relationship this becomes

$$\Delta S_1 = k \ln W = kN \ln \frac{N}{N - N_A} + kN_A \ln \frac{N - N_A}{V_A}$$

By using $L = L_0 + kT_E$ (L_0 is the heat of fusion, and L is the enthalpy of fusion), these equations reduce to

$$\frac{F_s}{NkT_E} + \frac{\alpha N_A (1 - N_A)}{N^2} - \ln N(N - N_A) - \frac{N_A}{N} \ln \frac{N - N_A}{N_A}$$

where

$$\alpha = \left(\frac{L_0}{kT_E}\right)\left(\frac{\eta_1}{v}\right)$$

This gives a relationship between the change of free energy F_s, the fraction of sites filled N_A/N, and the expression α. This, the variation of the free energy with the fraction of sites filled, can be plotted for various values of α. The resulting curve is given in Fig. 2.6.

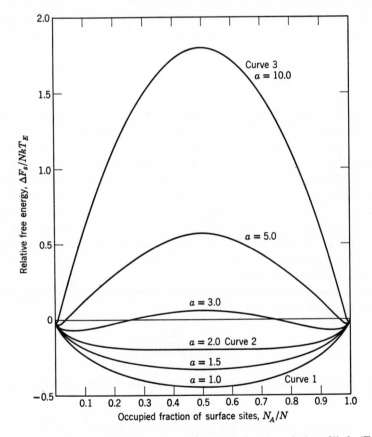

Fig. 2.6. Variation of free energy of interface with fraction of sites filled. (From Ref. 12, p. 181.)

It is apparent that the curve has a minimum at $N_A/N = 0.5$ for all values of α less than about 2, whereas for all higher values it has 2 minima, one at a very small value of N_A/N and the other for a value close to unity.

The minimum at 0.5 indicates that the free energy has a minimum value when half the sites are filled; that is, that a rough interface should be the equilibrium form. The 2 minima indicate that, for surfaces with appropriate values of α, we should expect either that very few sites are filled on a complete layer, or that almost all sites are filled, which corresponds to a few vacancies in a complete layer. These are 2 ways of describing a smooth interface, it being clear that a very small proportion of atoms on and vacant sites in an otherwise perfect layer is as close to perfection, or absolute smoothness, as it is possible to come except at the absolute zero. The reason is the relatively large decrease in free energy provided by the entropy contribution of a few atoms or vacancies that can be disposed in many different ways. The parameter α that controls the structure of the equilibrium interface consists of 2 parts: L_0/kT_E, which depends on the material, the crystal structure, and on the nature of the adjoining phase (whether melt, vapor, or solution), and η_1/v, which depends on the face under consideration. It is highest for the most closely packed planes, since it is the fraction of the total number of nearest neighbors that lie in the plane on which the atom rests; this is $6/12$ for a close-packed {111} plane, $4/12$ for a cube {100} plane, and smaller values for higher index planes.

The value of L/kT_E is less than 2 for all metals, for equilibrium between melt and crystal; typically, it is about 1.2, whereas for sublimation it is in the neighborhood of 20; since η_1/v is necessarily less than 0.5, it follows that α must be less than 2 for the melt-crystal interface for all metals, but it can be greater than 2 for the vapor-crystal interface. For organic crystals such as salol and glycerol, α is greater than 2 and these substances develop crystallographic faces, as distinct from metals, which do not.

The case of ice is an interesting one, as the values of α should be very different for surfaces parallel and perpendicular to the basal plane; the basal plane should have a value of α that is greater than 2, while all other planes should be rough. This is in agreement with experimental observations on the growth of ice crystals.

The conclusions that follow from Jackson's treatment of the interface are well supported by experiment except in one regard; if the interface were rough, in the sense of 50 per cent of the sites being filled, there would be some regions where all the sites were filled locally, forming a new completed layer, on which, in turn, half the sites would be filled, and so on. Thus, the single layer rough interface is not likely to exist; it would immediately become a multilayer interface and its mean inclination would quickly depend only on the local isotherms and not on the close-packed plane on which the structure was based. There is no evidence to contradict this conclusion, but as will be shown later, there is ample evidence that the growth rate in a given direction does depend on the atomic arrangement in planes perpendicular to that direction. This is difficult to understand if the surface is "ideally rough."

A different approach to the problem of the structure of the interface is that of Cahn (13) who concludes that it is not valid to discuss a surface as being singular or nonsingular, or to discuss whether growth is by the lateral propagation of steps or the forward motion of the interface without taking into account the effect of the driving force on the nature of the interface. Cahn proposes that the "step" between an atomic layer and an incomplete superimposed layer may be "sharp," in which case growth would be by lateral propagation of the step, or it may be "diffuse" to an extent that depends on the material and the driving force. In the extreme case of very diffuse steps, there are very many available trapping sites for atoms arriving from the liquid, and growth takes place by propagation of the interface normal to itself. This is typical of a high driving force; but it should be emphasized that the essence of this theory is that there is a continuous range of possibilities between completely smooth and completely rough, the criteria being both the material and the driving force.

Cahn shows that there is, for any interface, a driving force below which growth is by lateral propagation of steps, and above which normal propagation occurs. The value of the critical driving force depends on the diffuseness of the interface, being very large for a perfectly smooth interface, and decreasing as the roughness increases. The point of view developed by Cahn that the degree of diffuseness of the interface can vary over wide limits seems at first sight to be in conflict with that of Jackson, who shows that an interface is either rough or smooth; however, Jackson's analysis depends on the assumption that there is only a single layer that is partially filled, whereas Cahn's view is that the interface may extend through a considerably larger number of layers.

2.3 Equilibrium between a Pure Metal and Its Melt

When a surface of a crystal of a pure substance is in contact with its melt, there is a temperature above which melting takes place, and below which freezing occurs, that is, the crystal increases in size at the expense of the melt. This temperature is the *equilibrium temperature*, T_E. It is the temperature at which neither melting nor freezing takes place. The rate of melting or freezing is observable experimentally, and is proportional to the rate at which latent heat is supplied or removed; however, while a detailed study of this rate yields interesting results (14), it is a less powerful way of looking at these phenomena than that of Jackson and Chalmers (15, 16) who considered the net rate as the difference in the rates of two distinct processes, in one of which, melting, atoms leave the surface of the crystal and become part of the liquid, while in the other, freezing, atoms from the liquid become part of the solid.

In order to calculate these rates, it is necessary to make some assumptions, the first of which is that there is an "activated state" through which an atom must pass in either the freezing or the melting process. This activated state corresponds to an energy that may be higher, but cannot be lower, than the mean energy of the atoms in the liquid. It has so far also been necessary to assume that atoms arrive at or leave from sites of the "repeatable step" kind, that is, their change in energy between liquid and solid is equal to the latent heat per atom. This must be true on the average, but the assumption neglects the possibility that an atom may leave from a high energy (i.e., low coordination) site, having previously acquired part of its latent heat.

The unit process of melting or freezing is considered to occur when an atom, at the surface, satisfies three conditions simultaneously: (*a*) it must have sufficient energy to take it up to the activated state, (*b*) it must have a sufficiently large component of velocity normal to the interface, and (*c*) it must find a site in the other phase in which it can remain. We will now consider these three conditions.

(*a*) *The energy condition.* Let the cohesive energy per atom in the solid be E_S, and the energy of the activated state, i.e., the minimum energy required to make the transition from solid to liquid, E_A (Fig. 2.7).

The fraction of atoms at the interface at temperature T possessing an energy equal to or greater than E_A is $\exp\left[-(E_A - E_S)/kT\right]$ where

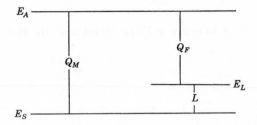

Fig. 2.7. Energies and activation energies for melting and freezing.

$E_A - E_S$ is the thermal energy required per atom. This can be written as exp $(-Q_M/RT)$, where Q_M is the activation energy for melting, in molar units. Similarly, the proportion of atoms in the liquid at the interface that have sufficient energy to make the transition is exp $[-(E_A - E_L)/kT]$ or exp $(-Q_F/RT)$. Since the latent heat of fusion L is the difference of the energy of the solid and the liquid, we can write $L = (E_L - E_S)$ or $(Q_M - Q_F)$ depending on the units used.

If the frequency of atomic vibrations is ν_S for the solid and ν_L for the liquid, and if the energy of each successive vibration is considered to be independent of that of the previous one, then the number of times per second that an atom at the interface has the required energy is ν_S exp $(-Q_M/RT)$ and ν_L exp $(-Q_P/RT)$ for melting and for freezing respectively.

(b) We will define geometric factors G_M and G_F as the proportion of atoms with sufficient energy that are moving in a direction with a sufficient component of velocity normal to the interface to escape.

(c) When an atom has sufficient energy and is moving in the right direction, it may find a suitable site at the surface of the other phase; this must be a site that is geometrically valid, and the atom must immediately lose some energy to its new neighbor by inelastic collision. If it fails to do so, it will instead rebound to the activated state and back to the phase it was in previously. The probability that an atom is accommodated in the other phase is called the accommodation coefficient A_M or A_F.

The number of times per second that all three conditions are satisfied will be $A_M G_M \nu_S$ exp $(-Q_M/RT)$, for melting, and $A_F G_F \nu_L$ exp $(-Q_F/RT)$ for freezing. If there are N_S and N_L atoms per unit area of the solid and liquid respectively at the interface, then the rates R_M and R_F for the two processes (expressed as the number of atoms crossing unit area per second), are:

$$R_M = N_S A_M G_M \nu_S \exp \left(-Q_M/RT \right)$$

and

$$R_F = N_L A_F G_F \nu_L \exp \left(-Q_F/RT \right)$$

The 2 processes ("melting" and "freezing") are regarded as occurring simultaneously and independently; their rates vary with temperature in the manner shown in Fig. 2.8. The curve for R_M rises more rapidly than that for R_F, corresponding to the higher activation energy for the melting process; R_M becomes larger than R_F, although it has a higher activation energy, because it has a higher accommodation coefficient; it is therefore not surprising that there is a temperature, the equilibrium temperature T_E, at which the rates are equal. Equilibrium therefore corresponds to a balance between the two competing processes, and not to a condition of "stagnation." A corollary to this conclusion is that the slightest departure in either direction from T_E is sufficient

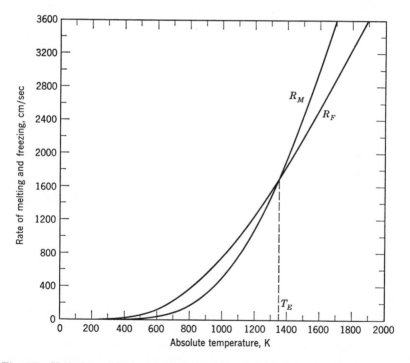

Fig. 2.8. Variation of the rates of the melting and freezing processes for copper with temperature. (From Ref. 16, p. 480.)

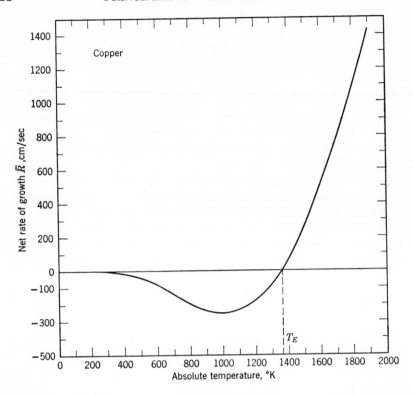

Fig. 2.9. Net rate of melting or freezing as a function of temperature. (From Ref. 16, p. 480.)

to upset the balance and to allow an excess of melting if the temperature is above T_E, or of freezing if it is below. The *net rate* $\bar{R}$ is the difference between R_F and R_M; it is given by the expression

$$\bar{R} = R_F - R_M$$
$$= N_L A_F G_S{}^{\nu_L} \exp\left(-Q_F/RT\right) - N_S A_M G_M{}^{\nu_S} \exp\left(-Q_M/RT\right)$$

and takes the form shown in Fig. 2.9. It should be pointed out that several of the parameters in this equation depend on the substance under consideration, and that both N_S and A_F may also vary with the crystallographic orientation of the interface; it follows that, although T_E is probably the same for all surface orientations of a given substance, the value of $\bar{R}$ for a given departure from equilibrium is certainly not constant.

At the equilibrium temperature T_E, the two rates are equal, or

$$N_S A_M G_M \nu_S \exp\left(-Q_M/RT_E\right) = N_L A_F G_F \nu_L \exp\left(-Q_F/RT_E\right)$$

or

$$\exp\left(\frac{Q_M - Q_F}{RT_E}\right) = \frac{A_M G_M N_S \nu_S}{A_F G_F N_L \nu_L}$$

but

$$Q_M - Q_F = L$$

hence

$$\frac{L}{RT_E} = \ln \frac{A_M G_M N_S \nu_S}{A_F G_F N_L \nu_L}$$

In the case of a plane solid-liquid interface, G_M and G_F are approximately equal, and N_S and N_L are also nearly equal; it will be assumed also that $\nu_S = \nu_L$, although Mott and Jones (17) have advanced reasons for believing this to be untrue. With these assumptions, we have

$$\frac{L}{RT_E} = \ln \frac{A_M}{A_F}$$

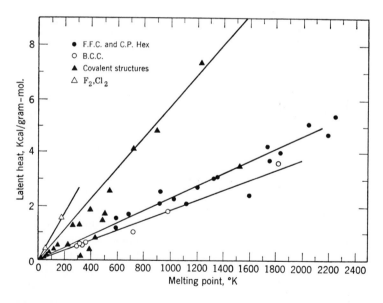

Fig. 2.10. Relationship between latent heat and melting point. (From B. Chalmers, *Physical Metallurgy*, John Wiley and Sons, New York, 1959, p. 85.)

Thus the equilibrium temperature is proportional to the latent heat of fusion, and also depends on the ratio of the two accommodation coefficients. It is possible that A_M is equal or close to unity, and that it is the same for all liquid metals, and perhaps for all liquids. The value of A_F would be expected to depend on the crystal structure of the solid, and on the crystallographic nature of the surface, if the structure of the solid surface has a recognizable relationship to any crystallographic plane. The experimental evidence shows that the value of T_E is proportional to L for a given crystal structure, and that it varies from one structure to another (16). Fig. 2.10.

It will be seen that the metallic structures (F.C.C., C.P.H., and B.C.C.) all have melting points that are high in relation to their latent heat compared with the structures which are covalently bonded; the difference is attributed to the much lower probability of an arriving atom reaching a site where it can satisfy the specific directional bonds of a covalent structure, while the less localized bonding of a metallic structure would be more tolerant. In the case of a molecular liquid such as fluorine, there is presumably an extra condition (for A_F) that the molecule must be correctly oriented in order to be accommodated. Some values are given in Table 2.1.

It will be noted that the expression for the equilibrium temperature T_E does not include E_A, the value of which is not known—but it may be estimated as follows. The energy barrier that an atom must surmount in order to move from solid to liquid or from liquid to solid may be identical with that encountered by an atom whenever it moves in the liquid, either by diffusion or by viscous flow; the activation energies for these processes, which are known in some cases, give an upper limit to the value of the energy barrier; part or all of the change in viscosity or diffusion with temperature may correspond to a change in the space available for the motion of atoms, in which case the height of the energy barrier would be less.

The actual values of the accommodation coefficients cannot be calculated from measurements relating to the equilibrium between solid and liquid, because the only parameters that can be determined at equilibrium are the latent heat and the equilibrium temperature, from which the ratio, but not the absolute values, of the accommodation coefficients can be found. Neither is it possible to calculate the accommodation coefficients from theoretical considerations without a knowledge of the height of the activation barrier over which atoms must pass during transitions between the crystal and the liquid; however, it follows from the definition of the accommodation coefficients that they cannot be greater than unity, and so the upper limit of the

Table 2.1.   Values of LA/T_E for Elements. LA is Latent Heat Times Atomic Weight

Structure	Element	T_E,°Abs.	L, Cal per Gram	LA/T_E
Face-centered cubic	Al	933	94	2.7
	Cu	1356	50.6	2.4
	Au	1436	16.1	2.3
	Pb	600	6.3	2.2
	Ni	1728	74	2.5
	Pd	1837	38	2.2
	Pt	2046	27	2.6
	Ag	1233	25	2.2
	Rh	2239	53	2.4
Close-packed hexagonal	Cd	594	13.2	2.5
	Mg	923	89	2.3
	Zn	692	24	2.3
Body-centered cubic	Ca	1123	52	1.8
	Cs	301	3.8	1.7
	Cr	2163	90.4	2.1
	Co	1768	63	2.1
	Fe	1912	65	1.9
	Li	459	100	1.5
	K	336	14.5	1.7
	Rb	312	6.1	1.7
	Na	311	27.5	1.7
	Sr	1043	25	2.1
	W	3683	44	2.0
Diamond cubic	Si	1703	396	6.5
	Ge	1232	102	6.0
Halogens	Br	266	16.2	4.9
	Cl	172	21.6	4.5
	F	50	10.1	3.8
	I	287	14.2	6.3
Other structures	Sn	505	14.5	3.4
	P	317	5	4.9
	Sb	903	38.3	5.1
	Bi	544	12.5	4.8
	Ga	303	19.2	4.4

From Ref. 16.

accommodation coefficient A_F can be calculated. It follows from the equation given above that, if $G_M = G_F$, and $\nu_S = \nu_L$, then $A_M/A_F = \exp(L/RT_E)$. The accommodation coefficient for melting cannot be greater than unity, hence, putting $A_M = 1$, as an upper limit,

$$A_F \lessgtr \frac{1}{\exp(L/RT)} = \frac{1}{3.16} = 0.32$$

A lower limit to the value of A_F can be deduced from the experimental observation that the temperature at which a mass of metal solidifies (as used, for example, in the calibration of a thermocouple) is not sensitive to ordinary variations in the rate of solidification. Freezing or melting point determinations on pure metals do not require extremely low rates of freezing or of melting; for example, if the freezing point of a metal is measured on a sample in the form of a cylinder 6 cm in diameter, it is common experience that the error introduced by allowing it to solidify in ten minutes would be negligible, say less than $\frac{1}{10}°$. This would correspond to a value of $d\bar{R}/dT \geq \frac{1}{20}$ cm/sec/degree. The value of $d\bar{R}/dT$ is computed as follows:

$$\bar{R} = N_S A_M G_M \nu_S \exp(-Q_M/RT) - N_L A_F G_F \nu_L \exp(-Q_F/RT)$$

whence

$$\frac{dR}{dT} = N_S A_M G_M \nu_S \frac{Q_M}{RT_E^2} \exp\left(\frac{-Q_M}{RT_E}\right) - N_L A_F G_F L \frac{Q_F}{RT_E^2} \exp\left(\frac{-Q_F}{RT_E}\right)$$

substituting $A_M/A_F = 3.16$, and assuming for copper that

$$G_M = G_F = \frac{1}{6}, \quad \nu_S = \nu_L = 2.8 \times 10^{13} \text{ sec}^{-1}, \quad N_S = \frac{(3 \times 10^{-8})^3}{\sqrt{2}}$$

$$\frac{Q_M}{RT_E} = 2.65, \quad \frac{Q_S}{RT_E} = 1.5, \quad T_E = 1356°$$

then

$$\frac{Q_F}{RT_E} = 1.15$$

then $(d\bar{R}/dT)_{T=T_E} = 8.4 A_M$ cm/sec/degree. If we adopt the assumed limiting value of $d\bar{R}/dT = \frac{1}{20}$ cm/sec/degree, then the lower limit of A_F for copper is found to be 0.002. The absolute values of the accommodation coefficients and the activation energies could be found, on the basis of the theory outlined above, if the relationship between R and T were determined experimentally. Orrok and Chalmers (18) attempted to measure ΔT directly during rapid melting of a lead crystal. The lead crystal contained a fine thermocouple, and fast melting was achieved by pressing a block of copper heated to about 800°C against the end of the crystal. The response of the thermocouple was recorded by means of an oscilloscope. The results were inconclusive in the sense that Orrok and Chalmers found that ΔT was less than about 0.1° for a rate $\bar{R}$ of 1 cm/sec. Thus the value of dR/dT is greater than 10.

If this value of $dR/dT = 10$ is inserted in the equation given above, a value of 0.3 is found for A_F; however, the existing evidence is not sufficiently definitive for a clear conclusion to be drawn as to whether all sites are available for trapping as would be implied by the value $A_F = 0.3$, or whether only a small fraction of possible sites are available, which would correspond to a lower value.

It will be recognized that the analysis given above has some limitations, the most serious of which is the implication that all sites at which atoms can leave the surface of the solid require equal energies, equal to the latent heat of fusion. It is possible that some of the atoms that actually leave the surface have higher energy, and therefore require less activation energy than the "repeatable step" atoms discussed above. This would mean that an atom acquires the total latent heat in two or more steps, the earlier ones corresponding to the departure of atoms from neighboring sites; another form of the same assumption is that all atoms at the surface can reach the activated state by acquiring thermal energy Q_M. The more likely situation, in fact, is that a much smaller number of atoms can reach the activated state with thermal energies that are much lower than Q_M, some perhaps as low as $Q_M - L/2$, corresponding to atoms with only three crystal neighbors.

Thus it is not to be expected that the equations derived above are quantitative accurate. They do, nevertheless, give a useful physical insight into the processes of melting and solidification, and in particular, into the significance of *equilibrium temperature, melting point,* and *freezing point.*

2.4   The Process of Crystal Growth

It is implicit in the foregoing discussion of the equilibrium between solids and liquids that single atoms arrive at and depart from the surface of the solid. It can easily be seen that almost if not quite the whole of the process of crystal growth must be by this single atom process. The only alternative that can be seen is that clusters or groups of atoms that have previously aggregated to form very small crystals should join together to form larger ones. The probability that a small crystal would be "in register" with the surface of a growing crystal is extremely small; it will be shown in Chapter 3 that the smaller a crystal is, the less stable it becomes; and it follows that the smallest possible groupings of atoms, which would be most likely to become correctly oriented when at the interface, would have an extremely short life.

The fact that individual crystals can easily be grown to any required size shows that the joining together during growth of crystals that are not correctly oriented must be extremely rare; but such an event would be thermodynamically only slightly less favorable, and much more probable, than joining with an exact alignment; and so it is again concluded that this process does not, in fact, contribute in any significant way to the growth of crystals from the melt.

Kinetics of crystal growth.   While it is clear that crystal growth proceeds by the addition of single atoms, and that the net observed process is the excess of atoms added over atoms lost, the details of the process must depend upon whether all sites on the surface are equally available for the addition of atoms and, similarly, whether all atoms are equally free to leave, or whether these processes can take place only at certain preferred sites, for example, at "steps" which form the edges of new layers. This question is, of course, very closely related to that of the atomic topography of the interface.

It has been shown that a useful criterion in terms of which this problem may be approached is the form of the relationship between the rate of solidification (or melting) $\bar{R}$ and the departure from equilibrium (ΔT). There are three distinct types of process to be considered, and they should lead to relationships between $\bar{R}$ and ΔT that are qualitatively as well as quantitatively distinct.

(a) It is predicted that, for an interface in which all sites are equivalent, $\bar{R}$ is proportional to ΔT. This follows, for small values of ΔT, from the equation, derived on page 42, for the value of $d\bar{R}/dT$, which contains the assumption that the accommodation coefficients are constant, that is, do not depend upon the value of ΔT.

(b) The second process is that in which the interface is atomically smooth, and new layers can form on it only by a nucleation process. In this case, the energy of an atom at a surface site depends very sharply on the number of its neighboring sites that are filled; the length of time that an individual atom is likely to remain in a site at which it has arrived increases greatly with each increase in its numbers of neighbors. The initiation of a new layer can take place, therefore, only when the rate of arrival is much higher than it would be for equilibrium, so that there is a high probability of a number of neighboring sites being filled simultaneously and thereby stabilizing each other. A detailed discussion of the "surface nucleation" problem is outside the scope of this book; details will be found elsewhere (19, 20). The conclusion that is reached is that the rate of growth should be related exponentially to the departure from equilibrium so that

$$\bar{R} = a \exp{(-b/\Delta T)}$$

(c) The third process is that in which atoms can arrive or leave only at steps, as in (b), but in which the steps are always present as a consequence of the existence of at least one dislocation whose Burgers vector has a component normal to the surface of the crystal.

The step, which is the only region at which atoms can be trapped, or from which they can escape, forms a spiral on the surface during growth, because all points on the step (except those very close to the point of emergence of the dislocation) advance at equal rates. It was shown by Hillig and Turnbull (21) that the distance between neighboring turns of the spiral is inversely proportional to ΔT, and therefore the total length of step is directly proportional to ΔT; the rate of growth, therefore, is proportional to $(\Delta T)^2$, because the rate of growth per unit *length of step* should also be proportional to ΔT.

A fourth case that should be considered is that in which the interface changes in character as ΔT is changed, as proposed by Cahn.

It would seem, therefore, that a direct experimental approach is possible to the problem of the nature of the interface and the type of growth; however, a serious difficulty arises in the measurement of the actual temperature of the interface while solidification is in progress. This is because the temperature differences are small, at least in the case of metals, and because a temperature gradient is necessary to provide for the extraction of latent heat. While the experiments of Orrok and Chalmers (18) referred to above give an indication of the value of $d\bar{R}/dT$, for a particular value of $\bar{R}$, they do not give any information on the form of the relationship between them; it is possible that this type of experiment could be refined to the point where it would resolve the question at issue, but a much more promising approach is provided by the recent work of Kramer and Tiller (22), who have shown theoretically that the relationship between the temperature of an interface and its speed of propagation can be obtained by measuring the thermal response of a crystal to a cyclic temperature fluctuation applied to a liquid in contact with it. The basis of the method is that if the interface between the solid and the liquid could absorb latent heat, and move in one direction, or emit latent heat and move the other way at exactly the same temperature, then the solid would be completely shielded from the temperature fluctuation. If, however, the temperature of the interface depends on the direction and speed of its motion (i.e., the sign and magnitude of $\bar{R}$), then a temperature wave will be transmitted into the solid. The method consists of the measurement and analysis of the thermal signal in the solid. Preliminary results (23) indicate that lead and tin behave according to the relationship

$$\bar{R} \propto (\Delta T)^2$$

for very small values of ΔT. This would indicate that, at least for very small driving forces, the interface is smooth and growth depends upon the screw dislocation mechanism. It is possible, and, as will be shown later, probable, that the screw dislocation mechanism is not significant at large driving forces.

There is a second kind of experimental evidence that strongly suggests that at very small driving force a metal may have a smooth interface. This is the observation of the "feather" type of growth that is sometimes found in ingots and other masses of metal that have solidified under small temperature gradients. The "feather" crystal, as is shown in Chapter 4 appears to grow by a smooth interface process in which the permanent step is provided not by a screw dislocation but by a twin boundary. On the other hand, the commonly observed cellular and dendritic types of growth strongly suggest that the interface must be rough at higher driving forces.

2.5   Solid-Liquid Equilibrium in Alloys

The kinetic description of the processes of melting and solidification of a pure substance can be extended to systems containing more than one chemical element (24). A crystal containing several species of atoms is in contact with a melt which contains the same species, not necessarily present at the same concentration; thus the component i is present at molar fractional concentration C_S^i in the solid and C_L^i in the liquid.

It is assumed that the melting and freezing processes are determined by the same expression as before, except that the rate for each species is proportional to its atomic concentration in the phase *from* which it moves. This implies the further assumption that the concentration in the surface layer of each phase is identical with that in the interior, i.e., with its bulk concentration.

Values of Q_M may in general be different for different species because a solute atom may be more or less tightly bound than the solvent atoms in a crystal. It will be assumed that Q_F is the same for all the species present, an assumption which, while almost cer-

tainly not strictly true, does not detract from the qualitative usefulness of the conclusions that can be drawn.

Then, for the species i, and for each other species present (including solvent as well as solute components)

$$R_M{}^i = A_M{}^i G_M{}^{\nu_S{}^i} N_S{}^i C_S{}^i \exp \left(-Q_M{}^i/RT\right)$$

and

$$R_F{}^i = A_F{}^i G_F{}^{\nu_L{}^i} N_L{}^i C_L{}^i \exp \left(-Q_F{}^i/RT\right)$$

The equilibrium condition for each component is given by

$$\frac{C_S{}^i N_S{}^i A_M{}^i G_M{}^{\nu_S{}^i}}{C_L{}^i N_L{}^i A_F{}^i G_F{}^{\nu_L{}^i}} = \exp\left[\frac{(Q_M{}^i - Q_F{}^i)}{RT}\right]$$

For equilibrium between the crystal and the melt, this equation must
be satisfied simultaneously for each component.

If it is assumed that the values of Q, A, G, and ν are independent of
concentration, and the general expression is applied to a binary solution we find that

$$n\,\frac{C_S{}^i}{C_L{}^i} = \frac{L^i}{R}\left(\frac{1}{T} - \frac{1}{T_E{}^i}\right)$$

which is a form of the Clausius Clapeyron equation. $T_E{}^i$ is the equilibrium temperature for the pure component i, and T that of the
solution.

If the two components are labeled A and B, then for dilute solutions
of B in A, $C_S{}^A \rightarrow 1$, $C_L{}^A \rightarrow 1$, and $T \rightarrow T_E{}^A$; then $\ln\,(C_S{}^A/C_L{}^A) \rightarrow$
$(C_L{}^B - C_L{}^A)$ calling $\Delta T^i = T_E{}^i - T$, we find that

$$\frac{L^A}{R(T_E{}^A)^2} = \frac{C_L{}^B}{\Delta T_A} - \frac{C_S{}^B}{\Delta T_A}$$

which is a form of the van t'Hoff relationship (25) giving the difference in slope of the liquidus and solidus lines at zero concentration
for any binary alloy in terms of the latent heat of fusion and the
melting point of the solvent phase. Neither the physical assumptions
nor the mathematical approximations leading to this expression are
valid except for dilute solutions; but this applies equally to the
thermodynamic derivation of the same expression.

Thus we find that equilibrium between a solid solution of one
composition and a liquid solution of another, at a temperature different
from the melting point of the solvent, is accounted for in terms of the
kinetic processes that take place at the interface; the basic assumption is that a solute atom in a crystal requires either more or less
energy to reach its active state than the atoms of the solvent. If

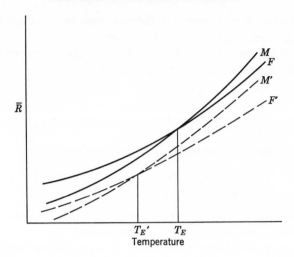

Fig. 2.11. Effect of a solute on the equilibrium temperature.

less, then it is easier for solute atoms to "melt" than for solvent atoms, while it is equally easy for both to "freeze" and consequently equilibrium will be established when there are fewer solute atoms in the solid than in the liquid. This permits equal rates of "freezing" and "melting" for both solvent and solute. The equilibrium temperature is also changed, for the following reason. Considering only the solvent, equilibrium is reached when equal numbers of solvent atoms go from solid to liquid and from liquid to solid; when both solid and liquid are pure, this is the equilibrium temperature. The rates of both processes are proportional to the concentration of solvent atoms, and in the case of an alloy this is lower in the liquid than in the solid (because the concentration of solute is higher). Thus the presence of the solute slows down the "freezing" process of the solvent more than the "melting" process and so equilibrium is reached at a lower temperature. This is shown in Fig. 2.11, in which M and F are the separate rate curves for the pure material, with an equilibrium temperature T_E, while the curves M' and F' are the curves for the solvent, F being decreased one and a half times as much as M. The new equilibrium temperature T_E' is necessarily lower than T_E. Thus the typical liquidus and solidus curves shown in Fig. 2.12a are accounted for qualitatively; in the opposite case, in which the solute atom has less energy than those of the solvent, and is more easily melted, the concentration of solute in the solid is greater, at equilibrium, than in the liquid and, by inversion of the previous argument, the equilibrium

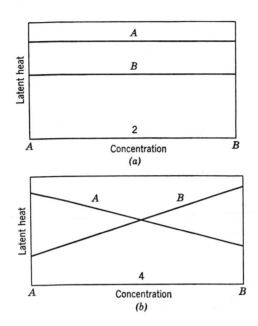

Fig. 2.14. Latent heat–concentration relationships for solid solutions. (From Ref. 24.)

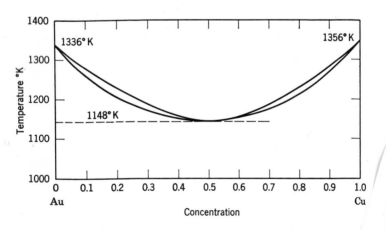

Fig. 2.15. Calculated phase diagram for the gold-copper system. (From R

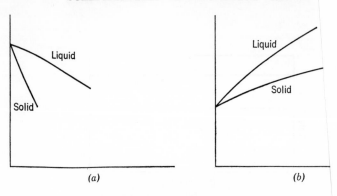

Fig. 2.12. Solidus-liquidus relationships.

temperature rises with increasing solute content. This is
Fig. 2.12b, which, although less common than the previ
represents a real situation.

The expressions derived above are based on the assumpt
the solid and liquid are both ideal solutions; this is alway
approximation for very dilute solutions, but it is often gr
correct for higher concentrations. Jackson (24) has shown
phase diagram for an alloy that conformed to ideal solution
at all concentrations would be as shown in Fig. 2.13, which i
approximation to the actual diagram for the gold-silver alloy

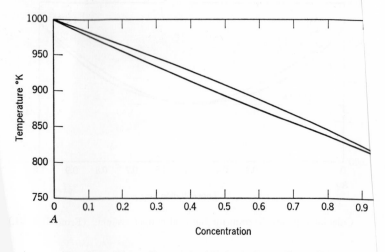

Fig. 2.13. Phase diagram for ideal solutions. (From Ref. 24.)

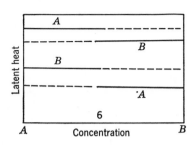

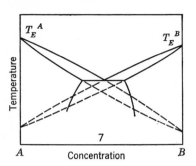

Fig. 2.16. Relationship between latent heat and concentration for an ideal eutectic forming system. (From Ref. 24.)

Fig. 2.17. Schematic phase diagram for eutectic forming system. (From Ref. 24.)

The assumption of ideality is equivalent to the case in which there is no interaction between the atoms, in the sense that each retains its own energy irrespective of its concentration. This is shown in Fig. 2.14a. The latent heat of the alloy can then be calculated directly from the latent heats of the pure components and the composition. If, however, the addition of one component tends to increase the energy of the other component, then the latent heat concentration relationship will be as shown in Fig. 2.14b. This gives a phase diagram of the type shown in Fig. 2.15, which corresponds closely to the experimental phase diagram for the gold-copper system.

The two cases considered so far correspond to complete mutual solid solubility. This requires that both pure components have the same crystal structure, and that certain other conditions are met. When the mutual solubility is limited, and two different terminal crystal structures are to be considered, the energy of each kind of atom in each structure must be taken into account. The relationship between the latent heat per atom and the concentration for the simplest case is shown in Fig. 2.16 for the system AB; the resulting phase diagram, shown in Fig. 2.17, will be recognized as a eutectic system, although it is somewhat idealized compared with most real cases.

2.6 Origin of Defects

No crystal is perfect in the sense that all the atom sites of a perfect lattice are filled with identical atoms; there are many kinds of imperfection, of which the most important are vacancies, interstitials, impurities, dislocations, cellular and lineage substructure, grain boundaries, and surfaces. The only type of imperfection that decreases the free energy of a crystal of a pure substance such as a metal is the vacancy, which is, therefore, always present in thermodynamic equilibrium. In a pure metal, the equilibrium concentration of interstitial atoms is so low that we may ignore them, and, as will be shown in Chapter 5, the cellular substructure is caused by impurities. A lineage boundary is a boundary between two regions of a crystal that differ in orientation, but which both grew from a common starting point. Thus it is possible to go from any point in a lineage structure to any other point without crossing a boundary, which is characterized by an abrupt change of orientation. At this point we are considering imperfections in crystals, and therefore grain boundaries are excluded. Since the subject of this section is the origin, rather than the consequences of imperfections, it is not necessary to pay further attention to surfaces, the origin of which is obvious. The imperfections to be considered in detail, therefore, are vacancies, dislocations, and lineages.

Vacancies. There is ample evidence, both theoretical and experimental, that some of the sites in any crystal are vacant. The theoretical evidence is based on a quantitative thermodynamic argument which corresponds to the concept that the increased entropy that results from the presence of some vacancies compensates for their energy, with the result that the free energy of the crystal is at a minimum if it contains a specific concentration of vacancies which depends on the temperature. This concentration is given by the relationship

$$\frac{n}{N} = \exp\left(-\frac{\Delta F}{kT}\right)$$

This concentration can be calculated as follows: let E_v be the energy required to create a vacancy (i.e., to move an atom from a site in the crystal to a site on its surface). Then if there are n vacancies in a crystal containing N atom sites, the change in free energy ΔF due to the vacancies is given (26) by

$$\Delta F = \Delta V - T\Delta S = nE_v - kT \ln \frac{N!}{(N-n)!n!}$$

in which the first term gives the increase in energy and the second the increase in entropy.

The equilibrium number of vacancies is that which minimizes the free energy, i.e., that for which

$$\left.\frac{\partial F}{\partial n}\right|_T = 0$$

but

$$\frac{\partial F}{\partial n}\bigg|_T = E_v - kT \ln \frac{N - n}{n}$$

and therefore, for equilibrium,

$$\ln \frac{n}{N - n} = \frac{E_v}{kT}$$

or

$$n \simeq \exp\left(-E_v/kT\right)$$

E_v is characteristically of the order of one electron volt, from which it follows that n/N at 1000° is of the order of 10^{-5}. This approximate result may be generalized in the form that the vacancy content of any metals at its melting point is of the order of 10^{-5}; the equilibrium vacancy content decreases rapidly as the temperature falls.

The experimental evidence for the existence of vacancies is largely based on diffusion, many aspects of which cannot be accounted for except in terms of the existence and motion of vacancies.

It is therefore of interest to inquire whether vacancies are "built in" as a crystal grows from the melt, and, if so, whether they are at their equilibrium concentration. It was pointed out by Chalmers (15) that vacancies could, in fact, be produced during solidification in the following way. The site X in Fig. 2.18 is to be filled by an atom, which

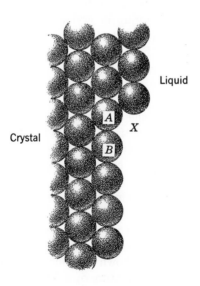

Fig. 2.18. Formation of a vacancy during solidification.

would normally come from the liquid; however, one of its neighbors, A or B, could fill it by a diffusive jump, creating a vacancy in the crystal. By comparing the activation energies for the diffusive jump Q_D and for the freezing process, Q_F, it can be seen that the probability p of creating a vacancy is given approximately by

$$p = \frac{\exp{(-Q_D/RT)}}{\exp{(-Q_P/RT)}}$$

which could have a value as high as 10^{-3}; thus one site in a thousand might be filled by the formation of a vacancy.

If the concentration of vacancies produced by the process of solidification is greater than that which would be in equilibrium, the question arises as to whether an excess vacancy concentration will be produced in the crystal, that is, whether the advance of the interface is fast enough for vacancies to be "left behind" in excess of the equilibrium number. The competing tendency is for the vacancies to reach the interface during their "random walk" diffusion. Doherty (27) has shown that the diffusive movement is so fast compared with normal rates of crystal growth that almost all the excess vacancies will be eliminated. Hence the "trapping" of a supersaturation of vacancies during solidification does not appear to be a real possibility, although there must be a few atomic layers of the crystal near the interface in which a high concentration may exist.

A much more sophisticated analysis by Webb (28) is based on his treatment of vacancy trapping in crystals grown from vapors. He finds that if the solid-liquid interface is considered to be diffuse, in the sense used above, and behaves as if it has a vacancy concentration of 0.1, then the concentration of trapped vacancies should be considerably in excess of the equilibrium value for the solid just below the melting point.

Since, as shown above, the equilibrium concentration of vacancies decreases as the temperature falls, it follows that supersaturation of vacancies must develop in a crystal if it cools down from the temperature at which it contains an equilibrium concentration of vacancies. Any region of the crystal is formed at a temperature very close to the solid-liquid equilibrium temperature, and then cools as the interface moves away from it.

There are three distinct ways in which the resulting supersaturation of vacancies can decrease: (a) grain boundaries and surfaces act as vacancy sinks; it has been shown by Doherty and Davis (29) that for aluminum a drop in temperature of a few degrees from a temperature near the melting point causes pits to appear on the surface except close to grain boundaries, including low angle sub-grain boundaries. This is interpreted to indicate that the boundaries are efficient sinks, in the sense that excess vacancies start "draining" into them as soon as cooling begins. This decreases the concentration of vacancies in the immediate vicinity of the boundaries. In regions that are more remote from the boundaries, the excess vacancies precipitate out in the form of surface pits, which, it is suspected, form where dislocations emerge.

Sinks of this, or any other, type can drain vacancies only from a limited distance, because the mechanism by which the vacancies reach the sink is by diffusion, a "random walk" process that takes place at a rate that is described by the diffusion coefficient D_v. The maximum distance, d, through which appreciable diffusion takes place in time, t, is given by $d \approx D_v^{1/2} t^{1/2}$, if the temperature is constant. If the temperature changes (as in the case under discussion) then the distance d is the integral of the distances traveled in successive elements of time, in each of which D_v has a different value. The value of d therefore depends on the rate of fall of temperature, but for crystals that grow under the conditions that usually prevail, d has a value between 0.01 cm and 0.1 cm. If this distance is substantially smaller than the effective "radius" of the crystal, many of the vacancies must remain within it.

(b) Dislocations that exist within the crystal may act as sinks; a vacancy can attach itself to an edge dislocation at a jog; this causes the jog to move along the dislocation by one interplanar space. Since this does not, in general, increase the energy of the dislocation, and it annihilates the vacancy, it follows that the free energy of the crystal is decreased. Hence this process, which causes climb of the dislocations, can persist, if there is sufficient time for diffusion, until the vacancy content is in equilibrium with the dislocations. A complication arises from the fact that, if a dislocation is initially straight between points at which it is pinned, climb requires an increase in its length. This slightly increases the vacancy content in equilibrium with the dislocation. If a dislocation is close to screw orientation, it has been demonstrated that it climbs to a helical form, to which further reference is made later.

(c) If surface, grain boundary, or dislocation sinks are not available within the diffusion distance imposed by the cooling rate, then vacancies may precipitate by aggregation into a cluster, which may take the form of a disc (either circular or related in shape to the symmetry of the crystal) or of a polyhedral void.

It follows from the foregoing remarks that the vacancy content of a crystal quickly reaches the new equilibrium value after a change of temperature, unless the temperature is so low that the vacancies are effectively immobile, in which case the main consequence of their presence, diffusion, is equally negligible.

Dislocations. The second type of defect to be considered is the dislocation. The relevant experimental facts are: that crystals of metals and of semiconductors grown from the melt usually contain dislocations; that the dislocation content depends upon the detailed conditions under which growth took place; and that under special conditions it is possible to grow crystals in which no dislocations can be detected by the methods that are currently available.

It is probable that there are a number of different ways in which dislocations may be introduced during crystal growth. Since dislocations are a nonequilibrium type of defect, they can be formed only as a result of nonequilibrium conditions during growth of the crystal; there are various kinds of disturbance that could be effective in producing dislocations. These may be classified as (a) externally applied stress of mechanical origin; (b) stresses of thermal origin; (c) local stresses due to concentration gradients; (d) the condensation of vacancies; (e) local stresses due to inclusions, and (f) "errors" in the growth process.

(a) EXTERNALLY APPLIED STRESSES. It is generally accepted that a small stress is sufficient to increase the dislocation content of a crystal to an enormous extent, but that the generation of dislocations where none were present cannot take place as a result of the application of a stress unless there is a very severe stress concentration. It therefore seems probable that if a growing crystal, either a single crystal or in a polycrystal, contains some dislocations, they may be multiplied as a result of stress, but it is equally clear that the stresses that are applied to a crystal during growth cannot generate dislocations if none were present initially.

(b) STRESSES OF THERMAL ORIGIN. There are many ways in which stresses may arise as a result of thermal expansion; for example, when a crystal is grown from a seed crystal (see Appendix) that contains some dislocations, the thermal shock that may occur when the cold seed is inserted into the hot melt may cause drastic multiplication of the dislocations that are present in the seed. Many of these dislocations will be propagated into the new crystal as it grows (30). In a somewhat similar way, the thermal shock that can occur when the growth of a crystal is terminated by withdrawing it from the melt can also cause a large increase in dislocation content (31). It is also believed that differential contraction of a crystal and the container in which it is growing, or between a crystal and its oxide (32) may set up stresses that are high enough to increase the dislocation content to a high value. It is also to be expected that high stresses will occur in a polycrystalline specimen of a material that has different thermal expansion characteristics in different crystallographic directions. The relief of these stresses by the motion and multiplication of dislocations cannot be avoided.

However, it does not appear to be possible that the dislocations that are normally present in carefully grown crystals can be caused by stresses of thermal origin, although some multiplication might occur of dislocations that are present as a result of some other mechanism.

(c) CONCENTRATION GRADIENTS. It will be shown later (Chapter 5) that an alloy or an impure metal usually solidifies with the solute distributed in a heterogeneous fashion. Tiller (33) has proposed a mechanism by which this could produce such high local concentration gradients that the resulting stresses would be relieved by the formation of dislocations. This mechanism is of doubtful validity, because it assumes that the crystal grows by the propagation of platelets or steps across its surface, and the theory that a metallic crystal grows by propagation across its surface of steps or platelets of optically visible thickness has been abandoned; the experimental evidence that was believed to support it (34) has been discredited by the more recent experiments of Chadwick (35) and of Weinberg (36) who showed that the surface steps produced by decanting the liquid from a growing crystal can, and probably do, arise during the solidification of the layer of liquid that adheres to the solid during decantation. The theoretical view, discussed on page 33, indicates that metals solidify by normal growth rather than by lateral propagation of steps. It therefore appears unlikely that dislocations are formed by concentration gradients unless either the conditions are such as to permit cellular or dendritic solidification (see Chapter 5) or changes of rate of solidification cause "banding" (see Chapter 5).

(d) CONDENSATION OF VACANCIES. It was originally suggested by Seitz (37) that a dislocation could be formed by the collapse of a disc of vacancies. The result of the collapse of a single layer of vacancies could be either a ring of partial dislocation surrounding a disc of stacking fault, or a ring of complete dislocation with no stacking fault. Price (38) has observed both of these types of dislocation in zinc, in crystals that had been cooled rapidly. Two questions therefore arise in connection with the possibility that dislocations are formed in this way in crystals grown from the melt. In the first place, does this process take place under the cooling conditions that are appropriate to the growth of crystals from the melt? And secondly, if they do, can their subsequent growth and motion produce the dislocation structures that are actually observed?

Jackson (39) has developed a theory for the formation of vacancy discs which leads to the conclusion that the supersaturation of vacancies should not be sufficient to cause discs to form spontaneously until the temperature had fallen far below the melting point (to 147°C for aluminum, and 382°C for copper). It has been shown by Tiller and Schoeck (40) that a vacancy disc should collapse to form a dislocation ring when its radius exceeds about 5 interatomic spacings; this critical radius is sensitive to the precise assumption that are made, and other estimates have differed substantially from this number. When a dislocation ring has formed, it can grow by climb, and, if it is a ring of complete dislocation, it can move by slip. Elbaum points out that the rate of growth should be very fast, because a disc does not form without very considerable supersaturation.* The dislocation loops may grow to such a size that they intersect each other, perhaps reacting to form the dislocation networks that are often found. Jackson's theory however, would suggest that vacancy discs form far too late in the process of cooling after growth for them to account for the dislocations that are often observed, and the fact that apparently dislocation free crystals have been grown (30) appears to support this view.

Howe and Elbaum (41) have shown that aluminum crystals, grown by the Czochralski technique, are apparently free from dislocations if they are less than about ½ mm thick; if they are thicker, they have an appreciable observable dislocation content. Howe and Elbaum computed that the "diffusion distance" for the conditions under which the crystals were grown was such that the surface should have been an effective sink for crystals up to about one-half millimeter in thickness, but that thicker crystals should retain a high proportion of the excess vacancies. This result supports the view that the dislocations originate by the condensation of vacancies to form discs; however, Elbaum has pointed out that the apparently dislocation free crystals might contain loops that are too small to detect by existing methods.

(e) STRESSES DUE TO INCLUSIONS. Jackson (42) has suggested a possible way out of the conflict between his theoretical conclusion and Elbaum's result; this is that dislocations might form with far less supersaturation of vacancies in a region of crystal that is imperfect, perhaps near inclusions, which usually exist in solid metals, and are

---

* This is an example of nucleation, which is discussed in detail in Chapter 3.

required to account for the experimentally observed nucleation characteristics (see Chapter 3). Large local stress fields must be set up around a particle of, for example, an oxide in a metal with which it is coherent, and it has been observed by transmission electron microscopy that dislocations exist around such inclusions. Jackson points out that the usual dislocation densities are compatible with the expected density of inclusions; this agreement is based, however, on "order of magnitude" estimates and makes no claim to accuracy.

(f) GROWTH ERRORS. The only other proposed mechanisms for the generation of dislocations as an inherent feature of growth are "growth errors," in which an embryo in the liquid becomes attached to the interface with an error of registry, thereby generating one or more dislocations (43), which would be "stretched out" and stabilized by the advance of the interface. Jackson (39) shows that this mechanism should not produce sufficient dislocations at ordinary rates of solidification. Another "growth error" mechanism (27) would be the production of a stacking fault which would relax either into vacancies or a dislocation. Here again the mechanism is possible, but cannot account quantitatively for the experimental results.

Lineage structures. It was suggested by Teghtsoonian and Chalmers (44, 45) and by Frank (46) that the lineage structures often observed in crystals grown from the melt are arrays of dislocations that originated by the condensation of vacancies and subsequent collapse of the discs to form dislocation loops. Since the lineage boundary extends from the interface back into the crystal, this mechanism would require that the loop should either have formed at the interface (i.e., a "half loop" with both ends of the dislocation "anchored" to, and growing with, the interface) or, if the disc nucleated behind the interface, it must grow, by climb, until it reaches the interface. It has been shown by Jackson (private communication) that the nucleation of a vacancy disc at the interface should occur far too infrequently to account for the observed lineage dislocation content, which is often 10^8 cm/cm^3; Schoeck and Tiller (40), on the other hand, have shown that a disc can never grow far enough by climb to catch up with the interface. It must therefore be concluded that, whether dislocation loops are formed by vacancy condensation or not, the dislocations that form the lineage type of structure must originate in some other, at present unknown, way. The most promising proposal, that of Jackson, is that they originate at solid particles included in the crystal, but in the absence of experimental support, this must be regarded as speculative.

In addition to the problem of the origin of the dislocations, the de-

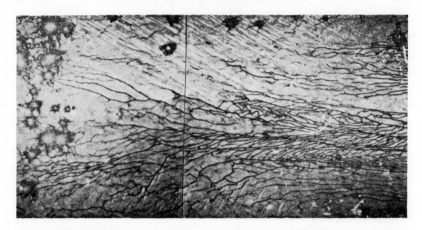

Fig. 2.19. Early stage in the formation of lineage structure in aluminum. Photograph by P. E. Doherty.

tailed morphology of the sub-boundaries requires explanation. For example, Doherty and Chalmers (47) have shown that the lineage structure in aluminum originates as a "river-type" structure, shown in Fig. 2.19. It is not clear why dislocations of like sign aggregate together when unlike dislocations could annihilate each other, or why similar low-angle boundaries join together instead of annihilating boundaries of opposite tilt.

References

1. F. C. Frank, *Proc. Roy. Soc.*, **215A,** 43 (1952).
2. J. D. Bernal, *Nature,* **183,** 141 (1959); **185,** 68 (1960).
3. F. Burton, N. Cabrera, and F. C. Frank, *Phil. Trans. Roy. Soc.*, **243,** 209 (1951)
4. F. C. Frank, *Discussions Faraday Soc.*, **5,** 48 (1958).
5. H. C. Gatos, P. L. Moody, and M. C. Lavine, *J.A.P.,* **31,** 312 (1960).
6. P. L. Moody, H. C. Gatos, and M. C. Lavine, *J.A.P.,* **31,** 1696 (1960).
7. G. R. Booker, *J.A.P.,* **33,** 750 (1962).
8. W. K. Burton and N. Cabrera, *Discussions Faraday Soc.*, **5,** 33 (1949).
9. W. W. Mullins, *Acta Met.,* **7,** 786 (1959).
10. K. A. Jackson, *Growth and Perfection of Crystals,* ed. Doremus, Roberts, and Turnbull, John Wiley and Sons, New York, 1958, p. 319.
11. K. A. Jackson, *Acta Met.,* **7,** 148 (1959).
12. K. A. Jackson, *Liquid Metals and Solidification,* American Society for Metals, Cleveland, 1958, p. 174.
13. J. W. Cahn, *Acta Met.,* **8,** 554 (1960).
14. D. Turnbull, *Thermodynamics in Physical Metallurgy,* American Society for Metals, Cleveland, p. 282.

15. B. Chalmers, *Trans. AIME,* 200, 519 (1954).
16. K. A. Jackson and B. Chalmers, *Can. J. Phys.,* 34, 473 (1956).
17. N. F. Mott and H. Jones, *Theory of the Properties of Metals and Alloys,* Clarendon Press, Oxford, 1936.
18. T. Orrok and B. Chalmers, Unpublished: Thesis, Harvard, 1958.
19. M. I. Volmer and M. Mander, *Z. Physik Chem.,* A 154, 97 (1931).
20. R. Kaischew and I. N. Stranskii, *Z. Physik Chem.,* 170, 295 (1934).
21. W. B. Hillig and D. Turnbull, *J. Chem. Phys.,* 24, 219 (1956).
22. J. J. Kramer and W. A. Tiller, *J. Chem. Phys.,* 37, 841 (1962).
23. W. A. Tiller, *Private Communication.*
24. K. A. Jackson, *Can. J. Phys.,* 36, 683 (1958).
25. A. J. C. Wilson, *J. Inst. Met.,* 70, 543 (1944).
26. C. Kittell, *Solid State Physics,* 2nd Ed., John Wiley & Sons, New York, 1956.
27. P. E. Doherty and B. Chalmers, Unpublished Thesis, Harvard, 1962.
28. W. W. Webb, *J.A.P.,* 33 (1961, 1962).
29. P. E. Doherty and R. S. Davis, *Acta Met.,* 7, 118 (1959).
30. W. C. Dash, *J.A.P.,* 29, 736 (1958).
31. R. S. Wagner, *J.A.P.,* 29, 1769 (1958).
32. C. Elbaum, *Progr. in Met. Phys.,* 8, 203 (1959).
33. W. A. Tiller, *J.A.P.,* 29, 611 (1958).
34. C. Elbaum and B. Chalmers, *Can. J. Phys.,* 33, 196 (1955).
35. G. Chadwick, *Acta Met.,* 10, 1 (1962).
36. F. Weinberg, *Trans. Met. Soc. AIME* 224, 628 (1962).
37. F. Seitz, *Imperfections in Nearly Perfect Crystals,* John Wiley & Sons, New York, 1952.
38. B. P. Price, *Phil. Mag.,* 5, 843 (1960).
39. K. A. Jackson, *Phil. Mag.,* 7, 1117 (1962).
40. G. Schoeck and W. A. Tiller, *Phil. Mag.,* 5, 43 (1960).
41. S. Howe and C. Elbaum, *J.A.P.,* 32, 742 (1961).
42. K. A. Jackson, *Phil. Mag.,* 7, 1615 (1962).
43. B. Chalmers, *Impurities and Imperfections,* American Society for Metals, Cleveland, 1955, p. 54.
44. E. Teghtsoonian and B. Chalmers, *Can. J. Phys.,* 29, 270 (1951).
45. E. Teghtsoonian and B. Chalmers, *Can. J. Phys.,* 30, 338 (1952).
46. F. C. Frank, "Deformation and Flow of Solids," IUTAM colloqium, Madrid, 1955, p. 3.
47. P. E. Doherty and B. Chalmers, *Trans. Met. Soc. AIME,* 224, 1124 (1962).

3

Nucleation

3.1 Metastability of Supercooled Liquids

The equilibrium considerations of Chapter 1 and the atomic mechanisms discussed in Chapter 2 imply that the temperature of a liquid cannot fall below the melting point, and, conversely, that of the solid cannot exceed it except by the amount required to "drive" the process of growth or of melting. In the case of metals, and of many other materials, this difference is quite small. It is, however, readily shown experimentally that a liquid can, in fact, exist for a long time at a temperature that is well below the melting point; for example, molten nickel can be cooled, under suitable conditions, to a temperature 250 degrees below the equilibrium temperature for nickel; it can be maintained as a liquid within this temperature range for an unlimited time. A condition that must be satisfied in order to *undercool* or *supercool* a liquid is that *none* of the solid be present. In order to understand the metastability of the supercooled liquid it is necessary to consider the first stages in the formation of a crystal.

At any temperature below that of equilibrium, the free energy of the solid is less than that of the liquid; this is a consequence of the thermodynamic definition of equilibrium. Transition from liquid to solid is therefore the direction in which a change can proceed spontaneously. Why, then, does not the supercooled liquid transform spontaneously to solid? The transformation of a liquid to a solid takes place, as discussed in Chapter 2, by the growth of crystals. It follows that when a solid begins to form in a molten material, it must at first take the form of extremely small crystals. The stability of a very small crystal differs from that of a large crystal because equilibrium between solid and liquid across a curved interface differs from that across a plane interface, which was implicitly assumed in the discussion of equilibrium in Chapters 1 and 2.

3.2 Equilibrium Conditions for a Curved Interface

The equilibrium between a solid and a liquid may be looked at in either thermodynamic or atomic terms; the effect of curvature of the interface on the equilibrium conditions may, in the same way, be considered from either point of view. We will first develop the atomic considerations, which enable us to grasp the physical significance of the curvature of the interface, and then proceed to the more general thermodynamic treatment.

Stability of a curved interface; atomic considerations. The discussion of the conditions for equilibrium between a solid and liquid (Chapter 2) contained the assumption that the geometrical factors for the two processes are equal. This is probably true so long as the interface is plane; if it is curved, then the geometrical factor for escape from the solid is changed; it increases if the surface is convex, and decreases if it is concave, because the available "escape angle" changes with curvature (1). The change in the geometrical factor may be assumed to be proportional to the curvature of the surface, that is, inversely proportional to its radius of curvature r. The radius to be associated with a "spherical" crystal containing n atoms is proportional to $n^{1/3}$. The geometrical factor for the inverse process (transition from liquid to solid) is probably changed in the reverse direction. The value of the geometrical factor $G_M{}^n$ for the solid to liquid transition for atoms at the surface of a spherical crystal containing n atoms (this is a convenient parameter for the curvature of the surface) is given (1) by

$$G_M{}^n = \left(1 + \frac{\alpha}{n^{1/3}}\right) G_M$$

and for the reverse process,

$$G_F{}^n = \left(1 - \frac{\alpha}{n^{1/3}}\right) G_F$$

It is also to be expected that the escape of atoms from a convex surface should be easier than from a flat surface because the atoms are, on the average, less firmly bound; i.e., they have, on the average, fewer nearest neighbors in the crystal. This effect depends on the radius of curvature, which can again be expressed in terms of the number of atoms contained in a sphere of the relevant radius. The average energy for escape is

$$L^n = \left(1 - \frac{\beta}{n^{1/3}}\right) L_1$$

where β is calculated from the number of neighbors, on the same basis as before. The value of β is 1.33 for the face-centered cubic structure.

It follows from these considerations that the rate for the melting process at any given temperature should increase with decrease of the radius of curvature of the surface of the crystal and that the rate of the freezing process should decrease. A corollary is that the temperature at which the rates of melting and freezing are equal should be reduced; that is, the temperature of equilibrium should be lower for a small crystal than for a larger one. Thus, at any temperature below T_E, there is a radius of curvature at which the rates of melting and of freezing are equal. This is the *critical radius r**. The solid and liquid separated by a surface with just the critical radius are in equilibrium, but the equilibrium is unstable because a slight rise of temperature causes the rate of melting to exceed the rate of freezing, with the result that part of the crystal melts, its radius decreases, and the departure from equilibrium increases; thus melting accelerates and the crystal disappears. Conversely, a slight drop in temperature causes growth and the consequent increase in radius raises the temperature of equilibrium and the rate of growth increases.

Thermodynamic treatment of equilibrium across a curved interface. There are several ways of considering equilibrium between a solid and a liquid separated by a curved interface. The most general treatment is that of Gibbs: It follows from the thermodynamic definition of equilibrium that if a solid phase and a liquid phase are in equilibrium, their free energies per unit quantity are equal. When they are separated by a *planar* interface, they are at the same pressure, and the temperature at which the free energies are equal when both are at a pressure of one atmosphere is the equilibrium temperature T_E. However, a pressure difference equal to $2\sigma/r$ exists across any interface of free energy σ per unit area and radius of curvature r, as can be seen by considering the work done by an excess pressure ΔP in a sphere of radius r (Fig. 3.1) when the radius is increased to $r + dr$. The work done is equal to the increase in free

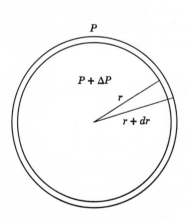

Fig. 3.1. Pressure due to surface tension.

P

$P + \Delta P$

r

$r + dr$

energy of the interface; thus

$$Pdr \times 4\pi r^2 = [4\pi(r + dr)^2 - 4\pi r^2]\sigma$$

whence

$$P = \frac{2\sigma}{r}$$

3.3 Calculation of Critical Radius

Equality of free energy, and therefore equilibrium, is achieved at any combination of temperature and radius at which the effect of this pressure difference ΔP on the free energy just compensates for the departure ΔT of the temperature from T_E. It can be shown that for the special case in which the solid is incompressible, i.e., in which no work is done in compressing the solid, ΔP is equal to the free energy difference that would exist between solid and liquid at $T - \Delta T$, *if the two phases were at the same pressure.* This difference of free energy, ΔG_p is therefore equal to ΔP; and

$$\Delta G_p = \frac{2\sigma}{r*}$$

The value of ΔG at constant pressure can be calculated as follows:

$$\Delta G_p = \Delta H - T\Delta S$$

at $T_E \Delta G_p = 0$; therefore $0 = \Delta H - T_E \Delta S$ or

$$\Delta S = \frac{\Delta H}{T_E}$$

$$\Delta G_p = \Delta H - \frac{T\Delta H}{T_E} = \frac{\Delta H(T_E - T)}{T_E} = \frac{L\Delta T}{T_E}$$

It follows that, for the special case of the incompressible solid,

$$\frac{L\Delta T}{T_E} = \frac{2\sigma}{r*}, \quad \text{or} \quad r* = \frac{2\sigma T_E}{L\Delta T}$$

It was assumed, for this derivation, that ΔH and ΔS are independent of temperature.

If the phase which is to be nucleated cannot be regarded as incompressible, then it is no longer valid to equate the free energy difference under equal pressures ΔG_p with the pressure difference ΔP. The expressions for the critical radius for a gas or a vapor nucleating in a liquid are, therefore, not the same as that given above.

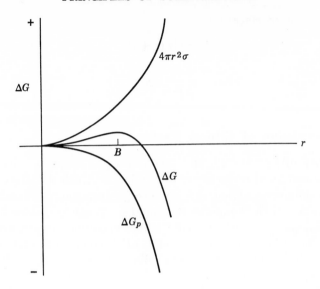

Fig. 3.2. Change of free energy due to the formation of a spherical crystal.

Another method which is often used for calculating the critical radius is the following: let ΔG_p again represent the difference of free energies of solid and liquid at $T_E - \Delta T$. It is implicitly assumed that the free energies of the two phases are independent of pressure; hence, as before,

$$\Delta G_p = \frac{L\Delta T}{T_E}$$

The change in free energy ΔG corresponding to a sphere of solid of radius r is, therefore, made up of a "volume" term, $\frac{4}{3}\pi r^3 \Delta G_p$, which is negative below T_E, and a "surface" term, $4\pi r^2\sigma$, which is always positive. Thus,

$$\Delta G = \tfrac{4}{3}\pi r^3 \Delta G_p + 4\pi r^2\sigma$$

The expressions for the volume term, the surface term and ΔG are plotted in Fig. 3.2, from which it will be seen that ΔG is positive for small values of r and negative for large values.

The critical radius r^* is defined as the radius, B, at which the derivative of free energy with respect to r is zero, that is, at which either a decrease or an increase in radius causes a decrease in the free energy of the system. The critical radius at a temperature $T - \Delta T$ is, therefore, the value of r for which

$$\frac{\partial G}{\partial r}\bigg|_{T} = 0$$

or,

$$0 = 4\pi r^2 \Delta G_p - 8\pi r \sigma$$

and

$$r^* = \frac{2\sigma}{\Delta G_p}$$

from which, using the expression given above for ΔG_p,

$$r^* = \frac{2\sigma T_E}{L\Delta T}$$

It follows from this expression, which came from both approaches to the calculation of the critical radius, that the relationship between r^* and ΔT takes the form shown in Fig. 3.3, in which the upper branch (for ΔT and r^* positive) corresponds to a crystal with a convex surface in a supercooled liquid, while the lower branch corresponds to a sphere of liquid in a crystal above its melting point (negative curvature, superheat). The numerical values of the critical radius corresponding to a given ΔT cannot be calculated unless σ is known; no method has yet been devised for measuring σ directly; its indirect determination is discussed on page 73.

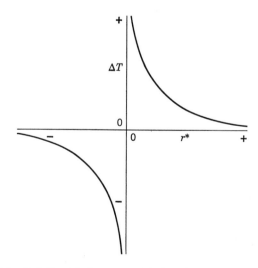

Fig. 3.3. Relationship between critical radius and supercooling.

A crystal that is below the critical size is called an *embryo;* if it reaches critical size, it becomes a *nucleus.* It should be recognized that although the thermodynamic considerations are most conveniently discussed in terms of a spherical crystal, the final result that defines the critical condition depends only on the curvature of the surface, and not on the actual existence of a complete sphere of crystal. Thus the thermodynamic and the kinetic considerations are in agreement in expressing the critical condition as a radius of curvature, although this conclusion implies the doubtful assumptions that the nucleus is so large compared with atomic dimensions that ordinary thermodynamics can be expected to apply, and that the curvature can be regarded as uniform, although, on an atomic scale, it cannot be so.

3.4 The Process of Nucleation

It was shown on page 23 that a liquid should, under equilibrium conditions, contain "clusters" of atoms, some of which have the structure of the corresponding solid. So long as the clusters are all below the critical size corresponding to the temperature of the liquid (i.e., embryos), they cannot grow to form crystals and no solid is formed. However, if the temperature is such that critical size is *less* than the largest cluster size, then nucleation occurs, and the supercritical clusters (nuclei) grow into crystals. The criterion that determines whether a liquid at a particular temperature will nucleate, or whether it will remain in metastable equilibrium is, therefore, the relationship between the size of the largest cluster and the critical size.

The number n_i of embryos of i atoms is, as shown on page 23, equal to

$$n \exp \left(- \Delta G/kT \right)$$

and

$$\Delta G = \frac{4}{3} \frac{\pi r^3}{V} \cdot \frac{\Delta H \Delta T}{T_E} + 4\pi r^2 \sigma$$

it follows that, since ΔG cannot be infinite, the ratio n_i/n cannot be zero. This means that there is always a finite probability of the existence of a cluster of any selected size; the problem of whether nucleation will occur or not in a given sample at a given temperature, therefore, becomes a question of the probability of an embryo reaching critical size in the sample (of limited size) in the time during which it is to be observed.

Theory of nucleation rate. It is therefore necessary to consider the rate of formation of critical sized clusters. The "classical" approach to this problem, originated by Volmer and Webber (2), for the condensation of vapors and applied by Turnbull and his co-workers (3) to solidification, is to calculate the rate of formation of nuclei on the basis that an embryo containing i atoms has a probability P_+ of capturing an atom, and becoming an "$i + 1$" cluster, and another probability P_- of losing an atom. In equilibrium, the values of n_i, n_{i+1}, etc., are such that the number of embryos leaving the "i" state in any given time is equal to the number joining it. The critical nucleus is the cluster for which $P_+ = P_-$ so that a nucleus of this size is as likely to grow as to melt.

If it is assumed that embryos of all sizes and all structures are in equilibrium, the number of critical nuclei $n_i{}^*$ per unit volume is given by

$$n_i{}^* = n \exp\left(-\frac{\Delta G^*}{kT}\right)$$

where G^* is the excess free energy of the critical nucleus, which is equal to $16\pi\sigma^3/(\Delta G_P)^2$. If it is now assumed that each critical nucleus grows into a crystal, and is thereby removed from the distribution of cluster sizes, the subsequent rate of formation of nuclei is determined by the rate at which smaller embryos reach critical size. The nucleation rate, I, is given by

$$I = ZS^*n_i{}^*$$

where Z is the net rate of transfer of atoms across the interface between the liquid and the embryo, S^* is the surface area of the critical nucleus, and $n_i{}^*$ is the equilibrium number of critical nuclei. This expression becomes

$$I = K_v \exp\left\{-[(\Delta G^* + \Delta G_A)kT]\right\}$$

where ΔG_A is the free energy of activation for the transfer of atoms from liquid to crystal, and K_v is given by

$$K_v = n^*(a\sigma/9\pi kT)^{1/2}n(kT/h)$$

where a depends on the *shape* of the nucleus (not necessarily spherical) n^* is the number of atoms in the surface of the critical nucleus, and n the number of atoms per unit volume of the liquid. This treatment of the nucleation rate assumes that there is always an equilibrium distribution of embryo sizes, although some of the largest embryos may be converted into nuclei, and grow to larger sizes. This assumption is sufficiently accurate for the analogous problem of nucleation of liquid drops in a vapor, but is not valid for the nucleation of crystals in a liquid, because of the much longer time taken for equilibrium to be re-established if it is disturbed, as a result of the lower mobility of the atoms in a liquid than in a vapor. The actual, or transient, nucleation rate, I_t, is given by

$$I = \frac{nkT}{h} \exp\left(-\frac{G_A}{kT}\right) \exp\left(-\frac{16\pi\sigma_{SL}{}^3 T_E{}^2}{3L^2(\Delta T)^2 kT}\right)$$

Nucleation temperature.   The form of the expression for I indicates that I remains very small until ΔT reaches a critical value, when it increases extremely fast, as shown in Fig. 3.4. In metallic and many other systems, this increase is so rapid that it is impossible experimentally to measure the nucleation rate; it is convenient to use the term "nucleation temperature" to designate the very narrow range between a rate that is too small to measure and one that is too large. It also follows from the expression for the nucleation rate that σ_{SL}, the interfacial free energy of the critical nucleus, could be calculated if the value of I were known for a given value of ΔT, since all the other quantities are known. The fact that ΔT and σ occur in the exponential indicates that the value of σ will be insensitive to the value of I; in fact, the calculated value of σ is not appreciably changed if I is assumed to be one per second or 100 per second. Hence if the nucleation temperature is measured, and a reasonable assumption is made for the value of I, σ can be calculated with an accuracy that is much more likely to be limited by the assumptions of the calculation than by uncertainty in the value of I.

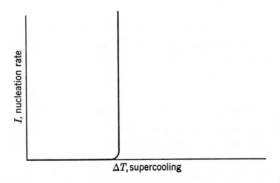

Fig. 3.4.   Nucleation rate as a function of supercooling.

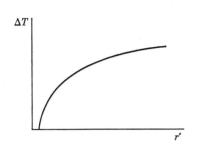

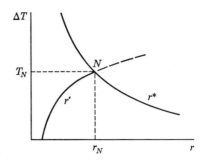

Fig. 3.5. Radius of largest embryo as a function of supercooling.

Fig. 3.6. The condition for nucleation.

A qualitative description of the significance of the nucleation temperature can be developed as follows: the maximum radius of embryo that is likely to be found increases with the undercooling, as shown schematically in Fig. 3.5. It is of interest to note that the maximum size is substantially greater than a single atom even at the melting point. This curve, however, does not have any absolute significance, because the maximum size, for a given temperature, should increase with larger samples and longer times. A combination of Figs. 3.3 and 3.5, to give Fig. 3.6, shows that the curves for critical radius and for maximum embryo size intersect at N. At this point, an embryo reaches critical size and becomes a nucleus. Therefore N represents the temperature and embryo size for nucleation.

Comparison between experiment and theory. Until relatively recently, it was believed that most metals could be supercooled only by rather small amounts, usually of the order of one degree, although exceptions had been observed. However, it was demonstrated by Turnbull and his coworkers (4) that when a sample of a metal is subdivided into a large number of very small drops, many of these can be supercooled by quite large amounts. The maximum supercooling that could be obtained for a given metal was shown to be very consistent even when the environment of the drop is changed, although some of the drops always nucleate before this limiting amount of supercooling is reached. Table 3.1 shows the values that have been obtained (5) experimentally for the limiting supercooling; the ratio of the supercooling to the melting point on the absolute scale is shown in Fig. 3.7.

Turnbull proposed that these results should be interpreted on the basis that the limiting supercooling is that which gives homogeneous

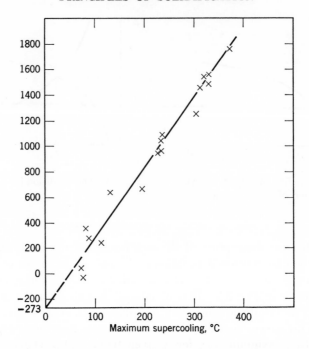

Fig. 3.7. Maximum supercooling as a function of melting point. (From *Thermodynamics in Physical Metallurgy*, American Society for Metals, Cleveland, 1911, p. 11.)

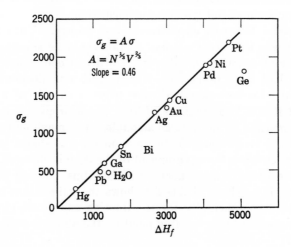

Fig. 3.8. Relationship between surface free energy and heat of fusion. (From Ref. 3, p. 294.)

nucleation, that is, nucleation by the process described above; nucleation with less supercooling is discussed on page 77.

If the maximum supercooling is taken as defining the temperature for homogeneous nucleation, as defined above, then the value of σ can be calculated from the theory (see page 70). Values of σ calculated in this way are given in Table 3.1 which also gives the ratio of the surface free energy per atom σ_A to the latent heat of fusion per atom H_F. See also Fig. 3.8.

Table 3.1.   Relationship between Maximum Supercooling, Solid-Liquid Interfacial Energy and Heat of Fusion[a]

Metal	Interfacial Energy σ (ergs/cm^2)	σ_g (cal/mole)	σ_g/L	ΔT_{MAX} (deg)
Mercury	24.4	296	0.53	77
Gallium	55.9	581	0.44	76
Tin	54.5	720	0.42	118
Bismuth	54.4	825	0.33	90
Lead	33.3	479	0.39	80
Antimony	101	1430	0.30	135
Germanium	181	2120	0.35	227
Silver	126	1240	0.46	227
Gold	132	1320	0.44	230
Copper	177	1360	0.44	236
Manganese	206	1660	0.48	308
Nickel	255	1860	0.44	319
Cobalt	234	1800	0.49	330
Iron	204	1580	0.45	295
Palladium	209	1850	0.45	332
Platinum	240	2140	0.45	370

[a] Data from D. Turnbull, *J. Appl. Phys.*, **21**, 1022 (1950) and Ref. 3.

It is significant that the ratio σ_A/L_A is roughly constant at a value of about 0.4. If the surface energy (which differs only slightly from the surface free energy) is calculated by the "nearest-neighbor bond" method discussed in Chapter 2, it is found that the value for a flat close-packed surface is $\sigma_A = 0.25L_A$ since each atom has 9 crystal neighbors out of a possible 12. However, it is evident that the energy per unit area should increase with increasing convexity of the surface, since each atom at the surface of a small crystal will have, on the average, fewer "crystal nearest neighbors," and therefore higher energy, than for the flat surface. It follows that for a crystal as small as the critical nucleus, the value of the surface energy per unit

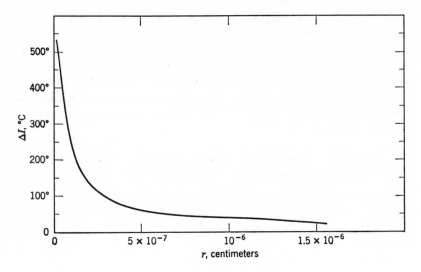

Fig. 3.9. Critical radius for copper as a function of supercooling.

area should be substantially greater than 0.25; the value of 0.4, derived from the nucleation experiments, is, therefore, in very good agreement with the theoretical estimate. This is strong additional support for the already overwhelming evidence that these are homogeneous nucleation events.

Size of the critical nucleus. Knowledge of the value of σ_{SL} allows us to calculate the actual size of the critical nucleus for any specified amount of undercooling. The following example is for copper, for which the latent heat of fusion, L, is 50 cal/gm, or 1.88×10^{10} ergs/cm^3; the melting point of copper is 1083°C, or 1356°K.

$$\sigma_{SL} = 1.44 \times 10^2 \text{ ergs/cm}^2$$

$$r^* = \frac{2\sigma T_E}{\Delta H \Delta T} = \frac{2 \cdot 54}{\Delta T} \times 10^{-5} \text{ cm}$$

This relationship is plotted in Fig. 3.9.

It is shown on page 41 that L is approximately proportional to T_E for the metals, and the experimental evidence quoted in Table 3.1 (5) shows that σ_g is proportional to L, and therefore to T_E; it follows that the value of A in the relationship

$$r^* = A \frac{T_E}{\Delta T}$$

should be approximately the same for all metals. A rather more precise general relationship between r^* and ΔT is provided by making the critical radius dimensionless by dividing it by a, the radius of the atom; then $r^*/a = \beta(T_E/\Delta T)$, which is shown in Fig. 3.10. For homogeneous nucleation of copper, the value of ΔT is 230°, which corresponds to a value of r^* equal to about 10^{-7} cm, which is less than 4 times the diameter of the copper atom. A spherical nucleus of this size would have a volume of about 4.2×10^{-21} cm³, and would contain about 360 atoms, since one atom of copper in a copper crystal occupies about 1.16×10^{-23} cm³. The corresponding values for other metals vary somewhat, but it is a good generalization to consider ΔT^* as about $0.2T_E$, and σ equal to about $0.4L$ when both are calculated on an atomic basis. Combining these approximate values gives a critical nucleus for homogeneous nucleation of about 200 atoms.

It is, however, not possible for a nucleus which has the crystal structure of copper to be exactly spherical in shape; if its radius is only about four atom diameters, the departure from an exact sphere must be substantial. This is also to be expected from the fact that the value of σ_{SL} may vary with the crystallographic nature of the surface which forms the solid-liquid interface. In view of the probable diffuse nature of the interface at very large supercooling, it is not realistic to specify the actual shape of the critical nucleus for homoge-

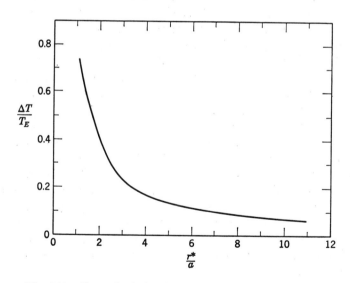

Fig. 3.10. General relationship between ΔT and r^* for metals.

neous nucleation; however, the crystal shown in Fig. 3.11 would be of approximately the right size, and would be bounded only by {111} and {100} planes. It is interesting to note that the faces of this crystal are close to their critical size for two dimensional nucleation at the critical temperature for the nucleus as a whole.

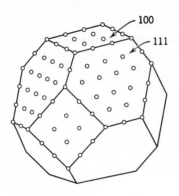

Fig. 3.11. Possible structure for the critical nucleus. (From B. Chalmers, *Physical Metallurgy,* John Wiley and Sons, New York, 1959, p. 246.)

3.5   Homogeneous Nucleation in Alloys

In the foregoing discussion, it was implicitly assumed that only a single kind of atom was present and, therefore, that the compositions of the embryo and of the nucleus are identical with that of the melt. If we now consider a binary alloy, it is apparent that the problem becomes more complicated, because the condition for unstable equilibrium that defines the critical temperature requires that the nucleus and the melt should be of different compositions; an extra variable is also introduced into the problem of the equilibrium distribution of embryo sizes. The theory of homogeneous nucleation in alloy melts is, therefore, complicated, and it has not been completely solved. There is, however, some experimental evidence (6, 7) that the nucleation of solid solution crystals from liquid solutions occurs at approximately the same supercooling, calculated from the liquidus temperature, as would be expected for a pure metal with the same melting point.

Recent work by Wagner (8), in the nucleation of solidification of various metals in binary germanium metal melts, has yielded very good confirmation of Turnbull's conclusion that homogeneous nucleation can be studied by the small drop technique. Some of Wagner's numerical results showed rather more undercooling than was found by Turnbull, but the differences are not large. Wagner was able to demonstrate that the germanium surface, on which the drop was supported, could be varied in orientation and perfection without changing the undercooling. The germanium, therefore, cannot have influenced the nucleation behavior of the metals.

3.6 Heterogeneous Nucleation

Metals in bulk usually nucleate at much less supercooling than the maximum observed in small drop experiments, and, as pointed out above, there are usually some drops which fail to supercool to the maximum extent. These results are consistent with the hypothesis, proposed by Turnbull (9), that the formation of a nucleus of critical size can be catalyzed by a suitable surface in contact with the liquid. The process is called "Heterogeneous Nucleation." The "nucleation catalyst" or "nucleant" may be either a solid particle suspended in the liquid, the surface of the container, or a solid film, such as oxide, on the surface of the liquid. One nucleation event is sufficient for the solidification of all the liquid that is in contact with the nucleation catalyst; but if a mass of liquid containing some nucleant particles is divided up into a sufficiently large number of small drops, most of them will not contain nucleants and will therefore not nucleate until the homogeneous nucleation temperature is reached. The most certain way, therefore, for achieving homogeneous nucleation is by subdivision; it is, however, possible to prepare "catalytically clean" bulk samples of some metals; Walker (10) has shown that nickel, iron, and cobalt can be so treated that they can be supercooled to the homogeneous nucleation temperature; the basis seems to be the high solubility of these metals, at temperatures well above their melting points, for oxide or other solid inclusions that would act as nucleants.

The physical basis for *heterogeneous nucleation* is as follows: it will be recalled that the condition for an embryo to become a nucleus is that its radius reaches the critical value for the amount of undercooling that is present. It is not, however, necessary for the embryo to be a complete sphere; any part of its surface that has a sufficiently large radius of curvature will have a better than even chance of growing. However, if one part of the surface of a solid has a radius of curvature that is larger than the average value, then there must also be regions with smaller radii; if the former grow, the latter may melt and the embryo will approach the spherical shape. Thus a non-spherical embryo will have a larger critical volume than a spherical one; and because departure from spherical form increases the surface area and therefore the surface free energy, per unit volume, it follows that the critical nucleus will approximate to a sphere, unless there is marked variation in the value of σ with the orientation of the surface.

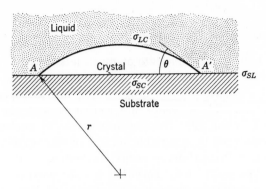

Fig. 3.12. Stability of an embryo on a substrate.

If, however, the embryo forms on a solid surface, as shown in Fig. 3.12, its stability depends upon the radius of curvature r, assumed to be uniform, and on the stability of the line of contact represented in the diagram by the points AA^1. The periphery of the embryo is stable if the total surface free energy of the system is unchanged by virtual displacement of the points A and A^1; this condition is satisfied if the horizontal components of the surface tensions balance, i.e., if $\sigma_{SL} - \sigma_{SC} = \sigma_{LC} \cos \theta$ where σ_{SL}, σ_{SC}, and σ_{LC} are the surface free energies of the interface between solid and liquid, solid and crystal, and liquid and crystal, respectively.

It is convenient to represent the surface energy relationship $(\sigma_{SL} - \sigma_{SC})/\sigma_{LC}$ by the parameter m, which, from the foregoing discussion, is equal to the cosine of the equilibrium contact angle. m is a measure of the tendency for the crystal to spread over the surface of the substrate; the term "wetting" is often used to describe this tendency, by analogy with the spreading of a liquid over the surface of a solid, which depends on an equivalent parameter. If $m > 1$, then there is no stable angle of contact, and the surface free energy is decreased continuously as the embryo spreads over the substrate; if $m < -1$, then again there is no stable angle of contact, and any contact between embryo and substrate causes an increase in surface free energy. Any value of m between $+1$ and -1 corresponds to a stable contact angle.

If m has a value approaching unity, then a nucleus of critical radius can form with a volume that is much less than would be necessary in the absence of the substrate. It should be emphasized that the critical radius r^* of the part of the nucleus in contact with the liquid is exactly the same whether a substrate is present or not; but a "cluster" containing a given number of atoms can form a spherical cap which has a much larger radius of curvature than a sphere of equal volume. A spherical cap can form on a substrate for which $0 < m < 1$. Thus the condition for nucleation—that the radius of curvature of the surface of the largest probable embryo is equal to the critical radius—can occur with less supercooling when a suitable substrate is present than without a substrate.

The volume of a spherical cap of height h (Fig. 3.13) is given by $\frac{1}{3}\pi h^2(3r - h)$ where r is the radius of the sphere. The ratio of the volume V_c of the cap to the volume V of the complete sphere of the same radius is

$$\frac{V_c}{V} = \frac{\frac{1}{3}\pi h^2(3r - h)}{\frac{4}{3}\pi r^3} = \frac{h^2}{4r^3}(3r - h)$$

If the ratio h/r is represented by q, then $V_c/V = q^2(3 - q)/4$. Thus V_s/V can be calculated for various values of h/r, with the following results; the corresponding values of m are also given.

h/r	V_s/V	m
1.	0.5	0.
0.8	0.35	0.21
0.5	0.16	0.50
0.3	0.06	0.72
0.1	0.007	0.9
0.01	0.00007	0.99
0.001	0.000007	0.999

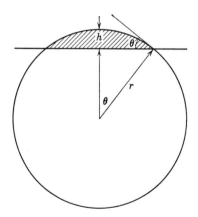

Fig. 3.13. Volume of a spherical cap.

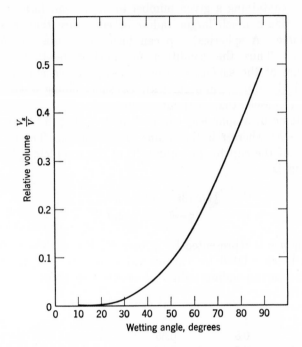

Fig. 3.14. Relation between volume of an embryo and angle of contact.

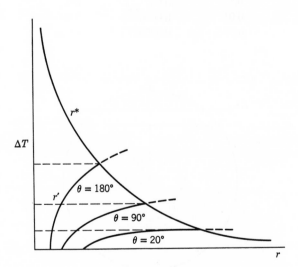

Fig. 3.15. Condition for heterogeneous nucleation (schematic).

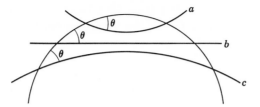

Fig. 3.16. Nucleation on curved (*a*), flat (*b*), and concave (*c*) surfaces, with the same angle of contact θ.

Thus the volume required for an embryo to reach a critical radius r^* depends on the angle of contact θ. The relationship between V_s/V and θ is shown in Fig. 3.14.

If it is assumed, and this is not excluded by the theoretical treatments, that the probability of occurrence of an embryo of a given size does not depend upon its shape, then it follows that, at a given undercooling, a larger radius is obtained (for the spherical cap of given volume) for smaller values of θ; the radius r' of the largest embryo is equal to the radius of the largest *spherical* embryo multiplied by the cube root of the ratio V_s/V. Figure 3.15 shows curves for r' for some values of θ. The intersections of these curves with the r^* curve give the condition for heterogeneous nucleation as a function of θ.

This treatment depends upon several assumptions; one is that the angle of contact is independent of temperature; it is also assumed that the area of the substrate is larger than the required area of contact of the embryo. Another hidden assumption is that the substrate is flat. If it is curved, on a scale comparable with the size of the embryo, then this would change the volume of embryo that would be bounded by the line of contact. This is shown in Fig. 3.16 in which 3 nuclei with equal radii and equal angles of contact are shown. It is evident that the nucleus has a smaller volume when the substrate is concave, and it therefore requires less undercooling to cause nucleation. The extreme form of a concave substrate is shown in Fig. 3.17;

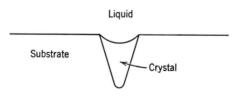

Fig. 3.17. Stability of a nucleus in a crack.

if the contact angle θ is less than 90°; i.e., if the interfacial energy of the solid-crystal interface is low compared with that of the solid-liquid interface, then the curvature of the interface can be negative, and the solid liquid-interface is stable even above the ordinary equilibrium temperature. If the geometry of a pit or crack in the substrate is such that an embryo of very small volume can form, nucleation should take place with very little supercooling.

A more general analysis that includes the areas of the interfaces has been developed by Fletcher (11), who writes

$$\Delta G = \Delta G_v V_2 + \sigma_{12} S_{12} + (\sigma_{23} - \sigma_{13}) S_{23}$$

where V_2 is the volume of the embryo, and the subscripts 1, 2, and 3 refer to the liquid, the embryo, and the substrate, respectively; σ refers to surface free energy per unit area, and S to the areas of the interfaces. Fletcher derives expressions and gives curves for the growth of embryos of various simple geometrical shapes on substrates which are also geometrically simple; his numerical conclusions relate to the nucleation of ice, but they could easily be modified to correspond to a metal.

If the substrate is rough on a scale that is small compared with the circle of contact, this has an effect on the nucleation process through the angle of contact. Effective nucleants have angles of contact θ less than 90°, that is, in which σ_{SC} is less than σ_{SL}. In such cases, roughness decreases θ and therefore makes the nucleant more effective. Let the roughness be represented by n, the ratio of the real to the projected area of contact; then the surface free energy equation becomes

$$n\sigma_{SC} - n\sigma_{SL} = \sigma_{LC} \cos \theta$$

or

$$\cos \theta = \frac{n(\sigma_{SC} - \sigma_{SL})}{\sigma_{LC}}$$

n is greater than 1, and therefore $\cos \theta$ is larger than it would be if n were 1, i.e., for a smooth surface. Thus θ is smaller for a rough surface than for a smooth one, and may become zero, in which case the undercooling for nucleation also becomes zero.

The effectiveness of a nucleant depends upon the extent to which its interface with the crystal is less energetic than its interface with the liquid. The theory of interphase interfaces is not sufficiently advanced to form a basis for quantitative prediction, but it is to be expected that a solid-solid interface should have low energy if the two-dimensional atomic lattices forming their matching surfaces are similar in geometry and in interatomic distance. Even if the match is good only in one dimension, some decrease of surface energy, compared with that of the substrate-liquid interface, is to be expected. Experience shows that this criterion is grossly inadequate as a basis for selection although it appears to provide a satisfactory explanation for the efficacy of many nucleants. The selection of nucleants is still on an empirical basis. Some of the known nucleants for metallic systems are listed in Chapter 8.

Sundquist (12) has pointed out that when the supercooling required for heterogeneous nucleation is small, the assumption that the critical nucleus is a spherical cap is no longer tenable. For example, if aluminum nucleates at one degree supercooling, the critical radius is about 800 atoms. The largest probable embryo at the melting point of aluminum must be smaller than it is at the temperature for homogeneous nucleation; we will assume it to contain 200 atoms. The best possible approximation to a spherical cap of radius 800 atoms containing 200 atoms would be a monolayer of about 8 atoms in radius. A monolayer of atoms on the surface of a substrate cannot be regarded as a group of atoms brought together by a fluctuation in the liquid; it is more reasonable to regard it as an adsorbed layer in which the atoms can be arranged in many ways. The groupings of the atoms in the adsorbed layer play the same part as the embryo in the liquid; some have the structure of the crystal and, if large enough, can provide the starting point for further growth. The calculation of the required size of two-dimensional embryos, and of the probability of their existence follow the same general lines as for homogeneous nucleation, but a modification is necessary because the number of atoms available to take part in the formation of embryos is not large compared to the number in the embryos, because only the adsorbed atoms can participate. Sundquist also points out that when the supercooling is small, as for example in the work of Glicksman and Childs (13) on tin, the conventional explanation would require a very improbable relationship between the three interfacial energies; for nucleation at supercooling of $4°$, $6°$, and $8°$, the difference of surface energies $(\sigma_{SL} - \sigma_{CL}/\sigma_{CL})$ would have to equal 0.999, 0.997 and 0.995, respectively. Sundquist suggests that very potent nucleants are much more likely to depend on a combination of shape and surface energy relationships than on the latter alone. This would also account for the failure of many proposed nucleants in which the interfacial free energy relationship is expected to be favorable.

3.7   The Nucleation of Melting

The equation for critical radius,

$$r^* = 2\sigma T_E / \Delta H \Delta T$$

is equally well satisfied for positive values of r^* and ΔT, as before, and for negative values of both variables. This would imply that a solid-liquid interface with negative curvature (center of curvature in the liquid) would be stable at temperatures above the melting point, and that it should be possible to superheat crystals above their melting points; there is evidence that this is possible under suitable conditions.

For example, Araki, Jackson, and Chalmers (14) have shown that ice, when internally heated, by focused infrared radiation or by radio frequency heating, superheats by a fraction of a degree before melting nucleates at discrete points; the existence of superheating was demonstrated by direct temperature measurements, and is confirmed by the "dendritic" morphology of the melted region, shown in Fig. 3.18. It has also been demonstrated by Turnbull, Mackenzie, and Ainslie (15) that substantial superheating can be achieved by the rapid heating of solids such as quartz, for which the kinetics of melting are slow; i.e., in

Fig. 3.18.   Negative dendritic crystal.

which the interface is at a temperature considerably above the equilibrium temperature during rapid melting.

Superheating of crystalline solids is, however, not usually observed because melting can, apparently, start at crystal surfaces without appreciable superheating. This is probably because the sum of solid-liquid and liquid-vapor interfacial free energies is less than the solid-vapor free energy, and there is, therefore, no increase of free energy in the early stage of melting. For example, in the case of gold, the values of the three interfacial free energies are:

solid-liquid	132 ergs/cm
liquid-vapor	1128 ergs/cm
solid-vapor	1400 ergs/cm

so that at the melting point the free energy is reduced by replacing a solid-liquid interface of a solid-liquid and a liquid-vapor interface, with a thin layer of liquid between. On the basis of numbers such as these, in fact, it is expected that a thin liquid layer should form on the surface below the melting point, because the difference in free energies $\sigma_{SV} - \sigma_{SL} + \sigma_{LV}$ could be used to convert solid to liquid. In the case of ice, there is some evidence for the existence of a liquid layer below the melting point, but similar evidence does not appear to exist for metals.

Nucleation of melting at the surface can be avoided by heating the interior to a higher temperature than the surface, or by providing the metal with a surface of negative curvature, i.e., a center of curvature outside the solid, and stabilizing the edges of means of a suitable substrate. This possibility was first pointed out by Turnbull (16) as a probable explanation of the fact, often observed experimentally, that if a single crystal of a metal or alloy is heated to a temperature a little above its melting point, and then cooled, it subsequently solidifies with the previous orientation; a polycrystalline specimen would revert to its previous grain size. It was thought at one time that these effects were due to the survival in the liquid of "crystallites" in some state intermediate between crystal and liquid. It is much more probable that the effect is, as proposed by Fisher, due to the superheating without melting of solid in fine cracks or crevasses in the walls of the container or in solid particles such as oxide fragments, the edges being stabilized by the crystal-liquid-substrate interfacial energies.

3.8 Dynamically Stimulated Nucleation

The preceding discussion of nucleation is based on observations and theory relating to situations in which the occurrence of nucleation is a function only of the temperature and the potency of any nucleants that are present. This condition is very well satisfied in liquids that are not in motion, but there are many observations which show that a variety of dynamic conditions in the liquid can cause the appearance of crystals that would not have been present in the corresponding static experiments.

There appear to be at least two main types of dynamically stimulated nucleation: that in which a completely metastable supercooled liquid (containing no crystals) is nucleated by a suitable dynamic event, and that in which the number of crystals in a solidifying liquid is greatly increased by dynamic means. The former type is certainly a nucleation process, but it cannot be stated with certainty that the latter is not due to the fragmentation of existing crystals. Dynamic nucleation is discussed here, and the formation of new crystals by the fragmentation of existing ones is considered in Chapter 8.

There are three distinct types of disturbance that have been shown to cause nucleation in a metastable melt—friction, vibration, and pressure pulse.

Nucleation by friction (for which the term "Tribonucleation" is proposed) is well known in the context of organic chemistry. It was first reported by Frankenheim (17), and frequently observed since, that a supercooled melt of an organic substance will often begin to solidify if the inside of the beaker containing it is "stroked" with the tip of a thermometer; a similar observation has been made with water supercooled by not less than about 0.5°. Two explanations have been proposed; one is based on the accepted view that friction between two surfaces that make contact is caused by the alternate welding and tearing apart of the asperities at which real contact occurs; the tearing away of material exposes new regions of surface whose potency as nucleation catalysts could be quite different from, and perhaps much superior to, that of a surface that has already been exposed to the melt. The only experimental evidence that relates directly to this point is that of Jackson and Chalmers (18) who demonstrated that although friction between two glass surfaces can cause nucleation of ice in water, breaking a glass rod in supercooled water does not cause nucleation. It can be concluded that the exposure of a new glass surface does not cause nucleation, and that this explanation must be rejected, at least for the nucleation of ice. The other explanation that has been proposed is that the vibration that is an inherent feature of the "stick-slip" type of friction has the same effect as vibration introduced by other means.

It has been shown by Walker (19) and by Jackson and Chalmers (18) that ultrasonic vibration of sufficient intensity causes nucleation in supercooled water. If, as a result of focusing, the maximum intensity of the vibration is within the liquid, nucleation takes place there; otherwise it takes place at an interface between the water and the containing vessel, or on the surface through which the vibration is transmitted to the water. There are indications that nucleation takes place more readily on surfaces that are poorly wetted by water (e.g., paraffin coated) than on those that are well wetted. It has been suggested that the potency of the surfaces in contact with the liquid could be increased by the cleaning action of the ultrasonic waves; but it has been shown (20) that ultrasonic irradiation immediately prior to supercooling does not cause nucleation. This explanation is also refuted by the fact that a single pressure pulse can cause nucleation. The more convincing explanation of the nucleation by vibration is that, as proposed by Walker (19), cavitation takes place during the negative pressure part of the cycle or series of cycles and that nucleation follows as a result either of the change in equilibrium temperature caused by the pressure changes during collapse of the cavitation bubble, or as a result of cooling of the surface of the bubble by evaporation during its growth (21). It should be pointed out that it has not been established that cavitation is a necessary condition for dynamic nucleation, but the conditions under which nucleation is observed appear to be generally similar to those in which cavitation is to be expected. However, there are indications that less ultrasonic or shock wave intensity is required for nucleation when the supercooling is larger than when it is small; this would not be predicted if the development of cavitation were in itself the criterion for nucleation. It would appear rather that cavitation must occur but is not effective unless either the bubble reaches a critical size before collapsing, or the bubble exists for a critical time.

We will now examine the two proposed mechanisms for nucleation by cavitation.

(1) Evaporation. It is well known that water can be cooled sufficiently rapidly by evaporation for ice to nucleate at its surface. The

classic illustration of this is the so-called "cryophorus," in which water is a tube containing only water and water vapor is made to evaporate very rapidly by cooling another part of the tube so that rapid condensation occurs there.

Nucleation can occur only if the temperature falls to the nucleation temperature (either homogeneous or heterogeneous) during the life of the bubble; it is therefore necessary to calculate first the rate of evaporation, and then the rate of fall of temperature at the surface of the bubble.

It has been shown (22) that, although the initial rate of fall of temperature at the surface of a cavity is high (23) (of the order of $1°$ in 5×10^{-5} sec) the total cooling that is possible is limited, in the case of water, to about 2 degrees, because of the increasing pressure of vapor in the bubble. It is therefore concluded that nucleation is unlikely to be caused by this mechanism.

(2) The alternative proposal (19) is that nucleation occurs at the existing temperature as a result of the displacement of the equilibrium temperature by the pressure pulse that occurs when a cavity collapses. The momentum of the liquid moving in to fill the cavity is sufficient to cause a very high pressure immediately after collapse; this would be followed by a very large negative pressure pulse. It follows from Le Chatelier's principle that the melting point of a substance that contracts on solidification is raised by an increase of pressure. The extent of this change can be calculated from the Clapeyron equation (see Chapter 1), from which it is found that a change in pressure of one atmosphere increases the melting point of nickel by $0.0027°$. Thus a change in equilibrium temperature of $100°$ would require 40,000 atmospheres change in pressure, a value considered to be reasonable. It should be pointed out that nearly all the metals decrease in volume on solidification, and their melting points therefore rise when the pressure increases. They can, therefore, in principle be nucleated by *increase* of pressure. On the other hand, silicon, germanium, gallium, bismuth, and water decrease in volume and would therefore require a large *decrease* in pressure to cause nucleation. There is ample evidence that water can be nucleated by ultrasonic stimulation, but there is no evidence in the case of the other substances with this characteristic.

It has been demonstrated by Walker (19) both for water and for nickel, cobalt, and iron, that a single pressure pulse can cause nucleation. A suitable pulse can be produced by dropping a steel ball onto the upper end of a fused silica rod of which the lower end dips into the supercooled liquid. It is found that the intensity of the pulse that is required depends on the amount of supercooling and on the characteristics of the end of the rod. The required intensity decreases with increased supercooling, and is significantly less for a fused silica rod with a broken end than for one that has been flame polished. The former characteristics may indicate that the size and duration of the cavities determine whether nucleation occurs or not; and the latter is compatible with the view that re-entrant corners in the end of the rod are conducive to cavitation.

The "single-pulse" observations, therefore, are compatible with the proposal that cavitation is a necessary, but not sufficient, condition for dynamic nucleation; the other condition, that less violent stimulation is required for as the supercooling increases, does not lead to a clear discrimination between the "Clapeyron" and the "Evaporation" mechanisms for nucleation.

It must be concluded that neither the evaporative cooling nor the "Clapeyron" theories for nucleation of cavitation is completely convincing, and that the former must be rejected on theoretical grounds. If it is believed that the dynamic effects change neither the relationship between critical size and undercooling, nor the kinetics of embryo building, then the most acceptable proposal is that the equilibrium temperature is displaced by the pressure pulse caused by collapse of the cavitation bubble.

3.9 Summary of Present Status of Nucleation Theory

As a result of the work of Turnbull and his coworkers, the theory of homogeneous nucleation of crystals in supercooled liquids is in a very satisfactory condition; the experimental results are in very close agreement with the predictions; the only remaining uncertainties are those that relate to the applicability of thermodynamics to very small crystals, and to the compatibility of the existing ideas on the structure of liquids with the presence of embryos in the thermodynamically calculated equilibrium distribution.

The general nature of heterogeneous nucleation appears to be well understood, but the detailed theory is less satisfactory, because it is not yet clear how the chemical, crystallographic, and geometrical characteristics of the nucleant determine its potency.

Dynamic nucleation, which has only recently been recognized as a separate problem, is very poorly understood, and it is not inconceivable that even the basic mechanism has not yet been uncovered.

References

1. K. A. Jackson and B. Chalmers, *Canad. J. Physics*, **34**, 473 (1956).
2. M. Volmer and A. Webber, *Z. Phys. Chem.*, **119**, 277 (1925).
3. For detailed references to Turnbull's many contributions, see J. H. Holloman and D. Turnbull, *Progr. in Met. Phys.*, **4**, 333 (1953).
4. D. Turnbull and R. E. Cech, *J.A.P.*, **21**, 804 (1950).
5. D. Turnbull, *J.A.P.*, **21**, 1022 (1950).
6. R. E. Cech and D. Turnbull, *J. Met.*, **3**, 242 (1951).
7. B. E. Sundquist and T. F. Mondolfo, *Trans. AIME*, **221**, 607 (1961).
8. R. S. Wagner, *AIME*, Conf. on Advanced Electronic Materials, **19**, 234 (1962).
9. D. Turnbull, *J. Chem. Phys.*, **20**, 411 (1952).
10. J. L. Walker, *Private Communication* (1961).
11. N. H. Fletcher, *J. Chem. Phys.*, **38**, 237 (1963).
12. B. Sundquist, *Acta Met.*, **11** (1963).
13. M. E. Glicksman and W. Childs, *Acta Met.* **10**, 925 (1962).
14. T. Araki, K. A. Jackson, and B. Chalmers, Unpublished work.
15. N. G. Ainslie, J. D. Mackenzie, and D. Turnbull, *J. Phys. Chem.*, **65**, 1718 (1961).
16. D. Turnbull, *J. Chem. Phys.*, **18**, 198 (1950).
17. P. Frankenheim, *Ann. Phys.* (Pogg), **111**, 1 (1960).
18. K. A. Jackson and B. Chalmers, Unpublished work.
19. J. L. Walker, *Private Communication*.
20. B. J. Mason and J. S. Maybank, *Quart. J. Roy. Meteor. Soc.*, **84**, 235 (1958).
21. B. Chalmers, Unpublished work.
22. R. Hickling, General Motors Conference, 1963; to be published.
23. B. Chalmers, General Motors Conference, 1963; to be published.

4

Microscopic Heat Flow
Considerations

4.1 Qualitative Observations

It has been shown in the preceding chapters that when a crystal is
in contact with its melt, the temperature of the interface determines
whether the quantity of crystalline material increases, decreases, or
remains stationary. If the interface is flat, then the temperature at
which neither solidification nor melting takes place is the equilibrium
temperature T_E. If the interface is curved, with radius r, there is still
an equilibrium temperature which, however, differs from T_E by an
amount ΔT which is given by $\Delta T = 2\sigma T_E/rL$. This is negative, i.e.,
the equilibrium temperature is lowered, if the center of curvature of
the interface is on the *solid* side of the interface. If the interface is
not at the equilibrium temperature, then either melting or solidification
occurs at a rate that increases with the difference between the actual
temperature and the equilibrium temperature. For small departures
from equilibrium, the rate is approximately proportional to the departure; however, the actual rate depends upon the crystallographic
orientation of the interface in a way that is known only qualitatively.
Although it is correct to regard the temperature given by

$$\Delta T = \frac{2\sigma T_E}{rL}$$

as one of equilibrium, in the sense that the opposing processes occur
at equal rates, it should be recalled that the equilibrium is unstable,
because a slight departure from the equilibrium temperature (say, in
the downward direction) causes freezing to occur, which increases the
radius and therefore, if the temperature remains unchanged, increases
the departure from equilibrium. It should be emphasized that the
foregoing remarks relate to the actual temperature of the interface

itself; this may be different from the temperature of the liquid or solid at even a short distance from the interface because of the latent heat of fusion that is generated at the interface during solidification, or is absorbed there during melting. At this point it will be assumed that the liquid is of uniform composition, and that, therefore, the equilibrium temperature is the same everywhere. The more complicated, but more usual, situation in which the composition is changed by the process of solidification will be discussed in Chapter 5.

When an interface is below the equilibrium temperature, and solidification therefore takes place, the latent heat that is evolved tends to decrease the supercooling. If the latent heat is not removed (by conduction) it very quickly eliminates the supercooling and suppresses the solidification process. The rate of removal of latent heat therefore controls the rate at which solidification can continue, and the interface temperature adjusts itself so that it corresponds to the rate of solidification determined by the externally imposed thermal conditions. The local rate of growth at any point on the surface therefore depends on the thermal conditions and on the orientation of the surface, since this influences the relationship between temperature and rate of growth. The interplay of the anisotropy of growth rate with the effects of the geometry of the surface on local heat flow is responsible for the very complicated morphology that may occur during solidification.

The earlier metallographers believed that metals characteristically solidified in the dendritic form first described in detail by Tschernoff (1), who apparently wrote about and reproduced drawings of dendrites in 1868. Because of the clarity with which it illustrates the meaning of "dendritic," as well as its historical significance, the well-known Tschernoff drawing is reproduced here.

With the recognition of the crystal structure of metals it was assumed, and later demonstrated, that the directions of the various dendrite arms coincided with crystallographically significant directions of the structure. More recently, some ambiguity has arisen in the use of the term "dendritic"; it is used here to indicate a linear, branched structure of which the arms are all parallel to specific crystallographic directions.

The first clear demonstration that solidification of a metal is not necessarily dendritic was that of Weinberg and Chalmers (2, 3), who showed that two quite distinct modes of crystal growth can occur in a "pure" liquid metal (lead); if there is a temperature gradient from liquid to solid, then the resulting interface shape is as one would predict from the shape of the isothermal surface that would exist in the

Fig. 4.1. Tschernoff's drawing of a dendrite. Reproduced from C. S. Smith, *A History of Metallography,* by permission of the Univ. of Chicago Press.

absence of solidification; the interface is smooth on a microscopic, but not necessarily atomic, scale, as can be demonstrated by decanting the liquid at any desired stage of the process. Figure 4.2 shows successive positions and shapes of the solid-liquid interface of a single crystal

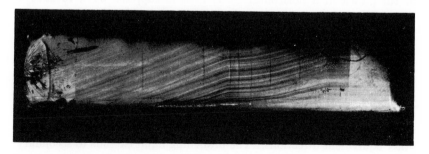

Fig. 4.2. Single crystal of tin, showing ripple marks made at intervals of one minute. (Note the lineage structure.)

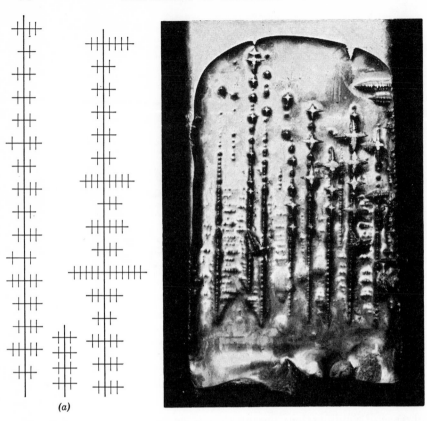

(a)

Fig. 4.3. Lead dendrites. (a) Schematic, (b) photograph. Photograph by F. Weinberg.

of tin grown in a horizontal boat; heat is supplied by a furnace at the right, and cooling takes place at the left. In this example the position of the interface was marked every minute by agitating the liquid so that a ripple ran along the surface and froze at the interface.

If, in an experiment similar to the one described above, the furnace is removed and the liquid loses heat fast enough to become supercooled, then dendritic solidification takes place. Dendritic solidification is characterized by a morphology resulting from the growth of long, thin spikes in specific crystallographic directions, with regular branches in other equivalent directions. The branching habit extends to secondary, tertiary, and sometimes higher orders (Fig. 4.3). In face-centered and body-centered cubic structures, dendritic growth is observed to take place in the cube directions, of which there are three that are mutually perpendicular. The main qualitative experimental observations are: (a) that dendritic growth takes place when, and only when, the melt is supercooled, (b) that the directions of growth are always strictly crystallographic, (c) branching occurs at roughly regular spacing, smaller for each successive order of branching, and (d) that only a small proportion of the liquid solidifies in this way; quantitative observations are referred to in the discussion of the theory.

4.2 Removal of Latent Heat

It has been shown that growth takes place at a rate that is determined by the difference ΔT between the actual temperature of the interface and the equilibrium temperature. Growth occurs at a constant rate; i.e., steady state conditions are achieved, when the rate of generation of latent heat is equal to the rate at which it is removed; the value of ΔT that corresponds to these conditions may therefore be regarded as a variable that is dependent on the rate of extraction of latent heat. For metals, the magnitude of ΔT is usually small compared to the externally imposed temperature differences, and it therefore may be ignored for the present purpose.

There are, in general, three ways in which the latent heat can be disposed of: (a) it can be conducted through the crystal into a heat sink, (b) it can be conducted away from the interface into the liquid, and (c) it can be absorbed as Peltier heat as a result of the thermoelectric effect that occurs at the interface between solid and liquid when a current is passed through the system in the appropriate direction. The third method will not be discussed in detail here because it is equivalent simply to changing the latent heat of fusion, and is of very little consequence for metals, as significant changes would require currents that are so large that the accompanying Joule heating would cause a far larger disturbance than the Peltier cooling. This is not the case for semiconductors, where the thermoelectric effects are much larger, and it has been demonstrated by O'Connor (4) that a substantial increase in the rate of steady state growth can be brought about by the use of Peltier cooling to remove latent heat from the interface. The discussion, therefore, is of the shape of a crystal growing: (a) when a temperature gradient is maintained in the crystal so that heat is conducted from the interface into the crystal, and (b) when the thermal conditions are such that heat flows from the interface into the liquid.

4.3   Extraction of Latent Heat by Conduction into the Crystal

It will be shown that when the supercooling required for growth is maintained by conduction of heat from the interface into the crystal, the interface advances uniformly and tends to be smooth, in the absence of the disturbing effects of solute that are discussed in Chapter 5. It will be assumed, in order to simplify the discussion, that the interface is initially planar, that it is isotropic in the sense that the relationship between V and ΔT is independent of the orientation of the surface, and that there is a heat sink in the form of a plane parallel to the interface maintained at a temperature below that of the interface. The geometry is shown in Fig. 4.4.

The stability of the planar interface is demonstrated as follows: Let the interface be perturbed, so that it takes the form shown in Fig. 4.5, in which region A of the interface is farther from the heat sink than the undisturbed parts. Since A is farther from SS than B is, the temperature gradient near A is less than that near B; less heat, therefore will flow through unit area of crystal near A than near B, and therefore the part of the crystal at A must grow more slowly than the part near B; the perturbation, therefore, will disappear and the planar interface will be restored.

This conclusion is not changed if heat is also conducted from the liquid, through the interface and through the crystal to the heat sink, in addition to the latent heat that originates at the interface; the temperature distribution would then be as shown in Fig. 4.6. The temperature gradient in the crystal is that required to carry $H + L$, where H is the heat reaching the interface by conduction in the liquid and L is the latent heat, i.e.,

$$\left.\frac{dT}{dx}\right|_c = \frac{H + L}{K_c}$$

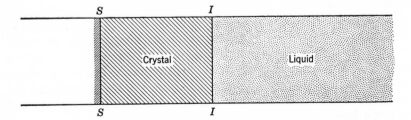

Fig. 4.4. Heat extraction through the solid. SS: a heat sink.

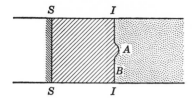

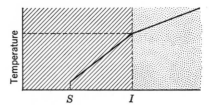

Fig. 4.5.   Stability of the interface.

Fig. 4.6.   Temperature distribution in liquid and solid.

where K_C is the thermal conductivity of the crystal.   For the liquid, ignoring convection,

$$\frac{dT}{dx}\bigg|_L = \frac{H}{K_L}$$

The contribution of the specific heat to the heat transfer equation has been ignored.   The value of K_C is usually higher than that of K_L; consequently the temperature gradient in the crystal may be equal to, greater than, or less than that in the liquid.

It can also be shown that the result arrived at above can be generalized to cases in which the interface is not planar; the interface tends to the configuration of the isothermal surfaces that would exist if no solidification were taking place, if the crystal is isotropic to thermal conduction; if it is anisotropic, the "isothermal" surface would be distorted.

The preceding discussion demonstrates that the rate of advance of the interface does not depend upon the temperature gradient in the liquid; by suitable adjustment of the supply of heat in the liquid and of the extraction of heat from the solid it is possible to vary the speed of advance without changing the gradient.   The independence of these two parameters has not always been recognized, perhaps because in very simple experimental systems, a change of rate is usually accompanied by a change in gradient.   However, the validity of the statement becomes obvious when it is recognized that an interface can advance, be stationary, or retreat while heat flows from the liquid to the solid.

Effect of surfaces and of grain boundaries.   The considerations of the preceding section relate implicitly to the growth of crystals in regions not affected by the proximity of external surfaces, such as the free surface of the melt or the surface of contact with the container, or internal surfaces such as grain boundaries.

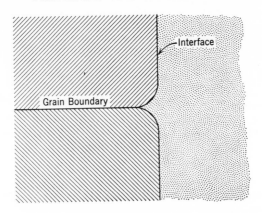

Fig. 4.7. Effect of grain boundary on interface shape.

It has been shown by Bolling and Tiller (5) that the shape of the interface should be disturbed in the vicinity of surfaces of these kinds; they have analyzed the shape of the solid-liquid interface near a grain boundary which lies in the solid in a plane perpendicular to the solid-liquid interface. They assume that thermodynamic equilibrium is maintained at all times. They conclude that a groove should exist at the boundary (Fig. 4.7), a result that can be derived from the static equilibrium of the intersection of two surfaces. They point out that the groove may be asymmetric if the surface free energies are not equal and that this can lead to the formation of grain boundaries in directions that are not normal to the solid-liquid interface. This phenomenon had been observed previously (6, 7) and accounted for in terms of the difference between the temperatures of different crystal faces growing at the same speed. This is another, not entirely unrelated, aspect of the anisotropy of solid-liquid interfaces. The inclination of the boundary leads to the selection during growth of crystals whose orientations are close to certain preferred values; this may lead to the development of ₂referred orientations if initially there are a large number of crystals of random orientation from which a smaller number are selected for survival. The origin of preferred orientation is discussed further in Chapter 8.

The effect of a free surface, also discussed by Bolling and Tiller, differs somewhat from that of a grain boundary, because the free surface is under less constraint. The morphology of the free surface is discussed in Chapter 8.

4.4   Conduction of Latent Heat Into the Liquid

It was pointed out above that if the liquid is at a lower temperature than the interface, dendritic growth is observed; this implies that an initially smooth interface is unstable in the presence of a temperature inversion.

The first question to be discussed is whether an *isotropic* surface would be stable when in contact with a supercooled melt; i.e., when there is a temperature inversion. This problem can be stated more precisely in the following form: if a solid with a spherical surface grows in an initially uniformly supercooled liquid (or a uniformly supersaturated solution), is the spherical shape stable? It was shown by Hamm (8) that, contrary to the intuitive conclusion, a sphere would grow as a sphere and an ellipsoid would grow as an ellipsoid of constant axial ratio. However, a more recent analysis by Mullins and Sekerka (9) shows that this is valid only if, (a) the sphere is perturbed only by the second harmonic (i.e., sphere to ellipsoid), and (b) the effect of curvature on the equilibrium condition is neglected. They show that the sphere is stable only if its radius of curvature is less than seven times the critical radius as defined above. The assumption underlying this analysis may not be applicable to the case of dendritic growth, however, because it assumes a sphere or ellipsoid growing in an infinite liquid initially at a uniform temperature; the more relevant case is one in which a planar interface, from which heat flows out into the liquid, is perturbed, as in Fig. 4.8a, by the addition of an ellipsoidal protuberance. The initial temperature distribution is shown in Fig. 4.8b. The protuberance, alone, would grow without change of axial ratio; but because it is growing in a temperature gradient, its axial ratio would increase (growth would be relatively slower near the plane, because the supercooling would be less there). The ellipsoid would therefore increase in length and become a spike. At the same time, the growth of the plane would be retarded near the base of the spike, because of the latent heat evolved by the spike itself; other spikes would form, at distances determined by the radius of the zone affected by the first ones; thus an array of spikes would grow. The lateral growth of each is retarded by the latent heat of the others and so the forward growth would predominate.

The occurrence of branching can be looked upon in much the same way; a primary arm, formed in the way described above, is itself in a region of temperature inversion. Uniform radial growth of the ap-

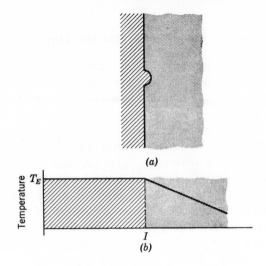

(a)

(b)

Fig. 4.8. Instability of interface with temperature inversion.

proximately cylindrical primary arm would be unstable, and a series of branches would occur, spaced at intervals determined by the latent heat evolved from the branches themselves. The same process can repeat on a finer and finer scale until there is not sufficient super-cooling to develop the irregularities of growth into branches.

It is implicit in the qualitative discussion of dendritic growth outlined above that the rate of growth at any given point is controlled by the rate of heat loss at that point. There are two respects in which this assumption should be examined more closely. In the first place, those substances that grow with a smooth interface can grow only by the lateral extension of existing steps, unless the supercooling is very large. If a crystal of such a substance is able to grow at moderate supercooling, it must do so by means of the perpetual steps provided by dislocations. In such cases, the faces will be nearly flat, since each

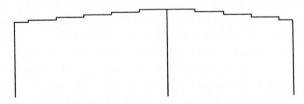

Fig. 4.9. Shape of surface resulting from growth by dislocation mechanism.

step travels outward to the limit set by the extent of the layer below; all points on the same face must grow at essentially the same speed. The faces will be slightly convex, as shown in Fig. 4.9.

Each face on which growth takes place (i.e., that translates perpendicularly to itself) requires at least one dislocation. It is difficult to envisage a mechanism that would supply dislocations for all the growing faces of a multiply branched dendrite. On the other hand, a crystal face that has a diffuse interface can grow at different rates at different points on the same face, which would be required if growth were controlled by the local heat flow. In particular, growth should, in this case, be greater near the corners and the edges than at the centers of faces. Thus it is concluded that dendritic growth, in the sense in which the expression is used here, is restricted to substances and growth conditions in which the solid-liquid interfaces are not completely smooth.

"Ribbon" crystals. The term "dendritic growth" is often used to describe a superficially similar, but actually somewhat different, type of growth that has been studied in crystals of silicon, germanium, and bismuth, each of which has a smooth interface under ordinary growth conditions. The crystals are grown by "pulling" at a controlled speed from a supercooled melt, and the result is a thin "ribbon-shaped" crystal, with its two flat surfaces parallel to a closely packed plane of the structure. In the case of materials with the diamond cube structure, the ribbon surfaces are {111} planes and the growth directions are ⟨112⟩. There is sometimes some branching at the edges of the ribbon, but this is always within the space bounded by the two planes which define the surfaces of the ribbon.

It has been shown by Wagner (10) that wherever ribbon growth occurs there are always at least two twin planes parallel to the ribbon surfaces. It can be seen that if a twinned crystal is bounded only by {111} planes, it will be of hexagonal shape, as shown in Fig. 4.10,

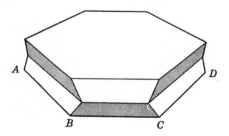

Fig. 4.10. Geometry of "ribbon" crystal.

and the twin plane will "crop out" as an "edge" which alternates between re-entrant (AB and CD) and exterior (BC).

It is proposed by Wagner that the re-entrant twin edge is a nucleation site for new layers on the edges of the sheet. If there are at lease two twin planes, growth in all directions is possible crystallographically, and growth actually takes place in the direction dictated by the imposed temperature gradient.

"Feather" growth. It is sometimes observed (11, 12, 13, 14) that some of the crystals in an aluminum ingot have a form best described as "feathering." These crystals, shown in Fig. 4.11, appear to resemble "ribbon" crystals in that they always contain twin planes, but it is not obvious why a twin mechanism should come into play in a material in which the interface is believed to be "diffuse." It is possible that the feather crystals grow only when the supercooling at the interface is so small that the diffuseness of the interface is insufficient to allow new layers to form: this explanation is not supported by the observation that "feather crystals" do not occur unless the conditions are such that growth is fast.

A somewhat different case of "flat dendrites" is that of ice, in which growth takes place dendritically in the basal plane of the structure,

Fig. 4.11. "Feather" crystals. [From Brenner and Roth, Z. *Metallkunde*, **32**, 10 (1940).]

Fig. 4.12. Dendritic morphology of ice growing in water at −1.5°C. Photograph by R. B. Williamson.

and much more slowly, and with a flat interface, in perpendicular directions. Figure 4.12 illustrates the type of dendrite which grows in water supercooled to −1.5°C. The difference between the growth habits in and normal to the basal plane is attributed to the existence of a smooth interface parallel to the basal plane and a diffuse interface at all other orientations.

4.5 Dendritic Growth

Four aspects of dendritic growth should be accounted for quantitatively in a satisfactory theoretical treatment; they are: (a) total amount of solid formed as a function of initial supercooling of the liquid, (b) speed of growth as a function of the temperature of the liquid, (c) direction of growth in relation to the structure of the growing crystal, and (d) spacing and relative lengths of the branches.

Total amount solidified. If it is assumed that the liquid is cooled to a uniform temperature (below that of equilibrium), isolated so that no heat may enter or leave it, and then nucleated, the total amount of solid that can be formed can be calculated from the latent heat of fusion, the specific heat and the amount of undercooling. Solidification by the dendritic process must cease when the amount of latent heat that has been evolved is sufficient to raise the temperature of the solid and the remaining liquid to the equilibrium temperature.

Let

L be the latent heat

C_S and C_L the specific heat of solid and liquid, and

ΔT the initial supercooling

then the fraction solidifying, S, is given by the equation

$$SL = C_L\left(1 - \frac{S}{2}\right)\Delta T + C_S\frac{S}{2}\Delta T$$

assuming that L, C_L, and C_S are independent of temperature; hence

$$S = \frac{2C_L\Delta T}{2L - (C_L - C_S)\Delta T}$$

or, if the two specific heats are assumed to be equal, and if ΔT is small, then $S \approx C\Delta T/L$.

In the case of lead, for which $C_L = 0.03$ cal/gm, $L = 6.26$ cal/gm, $S = 0.03/6.26 = 0.5\%$ for $\Delta T = 1$ degree. For the maximum possible supercooling, which for lead is about 80 degrees, only about 40 per cent of the liquid could solidify dendritically; the remaining liquid can solidify only by the extraction of heat from the solidifying material and, since this process would start at the outside, the heat would be removed, except at the very beginning, by conduction through the solid which had already formed. The solidification of this liquid would, therefore, be by the advance of a non-dendritic interface through the dendritic skeleton which had already formed. It was shown by Weinberg and Chalmers (3) that the "filling-in" stage is much slower than the dendritic growth.

Speed of growth. The problem of predicting the speed of growth as a function of undercooling has not yet been solved satisfactorily; the reason is that the rate of advance of the tip of a dendrite depends upon the rate at which heat is conducted from the tip into the surrounding liquid; this in turn depends on the shape and size of the tip and on its temperature. Jackson (15) has stated the problem in the following terms. The total supercooling of the ambient liquid may be regarded as being divided into three parts.

(*a*) The temperature difference between the interface (at any point of the surface) and the liquid remote from the interface.

(*b*) The difference in temperature between the interface and its equilibrium temperature.

(*c*) The difference between the equilibrium temperature of the tip and that for a planar interface.

Of these, (*a*) is the driving force for conduction of heat from the interface into the liquid; (*b*) is the driving force for the kinetics of the interface process, and (*c*) is a driving force that tends to increase the radius of curvature (or a constraint due to the surface energy). If there is a steady state rate of advance, it will occur when all three parts of the supercooling have constant values.

STEADY STATE THEORIES.   The first attempt to predict the steady state condition was that of Fisher (16) who made the following assumptions: (*a*) the tip is spherical and isotropic; (*b*) the kinetic driving force is zero; (*c*) the spherical tip loses heat by radial conduction, and yet moves forward with its shape unchanged.

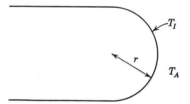

Let the tip be a hemisphere of radius r (Fig. 4.13), at a temperature T_I. The equilibrium temperature is T_E and the initial temperature of the liquid is T_A. Then it can

Fig. 4.13. Diagram for Fisher's theory of dendritic growth.

be shown that the rate of loss of heat H from the hemisphere at steady state is $H = 2\pi r K (T_I - T_A)$. But the rate of advance of the tip, v, is given by the volume solidified per unit time divided by the area of cross section, πr^2,

or

$$v = \frac{H}{L \rho \pi r^2}$$

hence

$$v = \frac{2K}{L \rho r} (T_I - T_A)$$

If the interface kinetic term is zero, then the radius of the interface must be that required for equilibrium between the solid and the liquid;

i.e., it must be the critical radius for the actual interface temperature; therefore,

$$r = \frac{2\sigma T_E}{L(T_E - T_I)}$$

Using this expression for r, the value of v is given by

$$v = \frac{2KL(T_E - T_I)}{2L\rho\sigma T_E}\,(T_I - T_A)$$

$$= \frac{K}{\rho\sigma T_E}\,(T_E - T_I)(T_I - T_A)$$

which cannot be solved for v without further assumptions because T_I is unknown.

It is concluded that, within the assumptions on which this analysis is based, the condition that the rate of production of latent heat is balanced by its rate of conduction outward can be satisfied for any value of interface temperature between T_E and T_I; the radius of the tip and the velocity vary accordingly. Fisher assumed that the correct solution would correspond to the maximum value of v, which occurs when $(T_E - T_I) = (T_I - T_A)$, and is $v = K(\Delta T)^2/4\rho\sigma T_E$, where ΔT is the supercooling of the liquid in which growth takes place; i.e., $T_E - T_A$.

The experimental results, quoted below, indicate rough agreement with the relationship

$$v \propto (\Delta T)^2$$

however, the quantitative agreement is poor; for example, for tin, v should be about $7(\Delta T)^2$, but experimentally it is about $0.1\ (\Delta T)^2$.

Examination of Fisher's analysis shows that there are at least four ways in which it could be modified without departing from the physical basis of the theory: there are, (a) to select a shape that can propagate without change, (b) to recognize that the surface may not be isothermal, (c) to introduce a term for the kinetic driving force, and (d) to include the possibility that the kinetic rate constant depends on the orientation of the surface.

It was pointed out by Chalmers and Jackson (17) that the observed departure from the $v \propto (\Delta T)^2$ law could be largely resolved by including a kinetic term, whose magnitude was an adjustable parameter; however, the required kinetic term would be unreasonably large, and it was concluded that this was not a sufficient refinement of Fisher's theory to be regarded as a solution of the problem; they also pointed out that the spherical shape assumed by Fisher was unrealistic, because a hemisphere, growing outward by uniform loss of heat, could not move forward without increase of radius, as assumed by Fisher. They also pointed out that a more "pointed" shape, which might solve this problem, would also improve the quantitative agreement with experiment by allowing the radius that is effective for heat flow to be larger than the minimum radius determined by the surface energy constraint.

It was shown by Horvay and Cahn (18) that an elliptical paraboloid with an isothermal surface would advance under steady state conditions (i.e., at a constant velocity and constant shape), and that the velocity v would depend on the supercooling ΔT according to $v \propto (\Delta T)^n/r$, where r is the tip radius. The exponent n, previously believed to have the value 1, was found by Horvay and Cahn to vary from about 1.2, for a paraboloidal dendrite of circular section, to 2 for a dendritic platelet. However, Horvay and Cahn reiterate the conclusion referred to above, that the heat transfer equation alone does not provide sufficient information to lead to a prediction of the actual velocity as a function of ΔT. It is also pointed out in the same paper that the assumption of uniform surface temperature is not realistic; they conclude, however, that this would require only a small correction.

Bolling and Tiller (19) have made essentially similar calculations, and have shown how to include a correction for nonuniformity of the surface temperature. They show that if the condition of uniformity of surface temperature is relaxed, then the stable shape would change slightly from a paraboloid. The correction, however, appears to be small. In the development of their theory, Bolling and Tiller follow the physical basis proposed by Fisher, stating the velocity criterion as follows: "the dendrite will develop those particular body dimensions that allow it to satisfy all the constraints, and which, simultaneously, allow it to propagate with the maximum possible velocity for a particular value of ΔT." No justification is attempted for the use of the maximum velocity criterion, except that it allows reasonable fit between "theory" and experiment. It could be argued that the necessity to introduce an arbitrary criterion of this kind implies an inadequate physical understanding of the process. A slightly different approach to the solution of the heat-flow problem is that of Timken (20) whose conclusions, however, are not very different from those of Horvay and Cahn, and of Bolling and Tiller.

NON-STEADY STATE THEORY. The various theoretical studies outlined above were based on the assumption that the growth of a dendrite is a steady state phenomenon; it is probable that true steady state conditions can be achieved in the growth of a silicon ribbon crystal, for example, and that the relationship between the various parameters can be analyzed realistically in terms of heat flow from the advancing "tip" or "edge." On the other hand, the periodic occurrence of branching in dendrites of metals or of ice suggests that the size of the tip and the temperature distribution around it may fluctuate in a periodic manner. The conclusions of Mullins and Sekerka (9) on the dependence of the stability of a curved interface on its radius of curvature indicate, as suggested by Cahn (21), that the tip should grow until it becomes large enough to be unstable, and then break down into a number of separate tips, each of smaller radius. Each of these tips again grows until it becomes unstable. The anisotropy of the surface free energy and of the growth kinetics should cause the instability to develop in a predictable, rather than a random, fashion, but it has yet to be demonstrated that the directions in which the "new" points form are those in which dendritic branching takes place, although this is an attractive hypothesis. The proposed process is illustrated in Fig. 4.14.

If the growth and branching of a dendrite are in fact two different aspects of the periodicity of the process, then it is not to be expected that a calculation based on steady state assumptions will be correct in detail; but since the over-all control of the process is still by heat flow into a colder liquid, the general form of the relationship should be unchanged.

It should be pointed out that the theories discussed above are all developed for a single isolated dendritic spike growing into liquid whose temperature is disturbed from its original condition of uniformity only by the heat conducted outward from the dendrite. This condition is

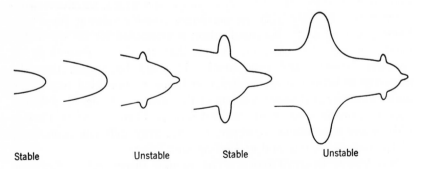

Stable   Unstable   Stable   Unstable

Fig. 4.14. Branching of dendrites.

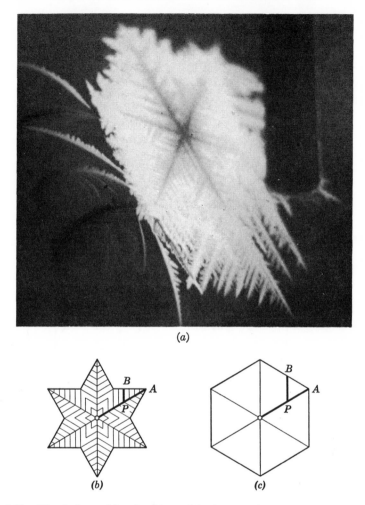

Fig. 4.15. Morphology of ice dendrites; (a) photograph of a dendrite, (b) schematic, (c) ideal. Photograph by R. B. Williamson.

not always satisfied, and it can be demonstrated that the secondary branches from the same primary branch, for example, do not grow independently. Each is influenced by the thermal field of its neighbors. The clearest illustration of this fact is seen in the characteristic morphology of ice dendrites growing in supercooled water; the observed morphology is shown in Fig. 4.15a, and is represented schematically in Fig. 4.15b. It is evident that the length PB of a secondary arm is less than that of the primary arm PA that originated at the same

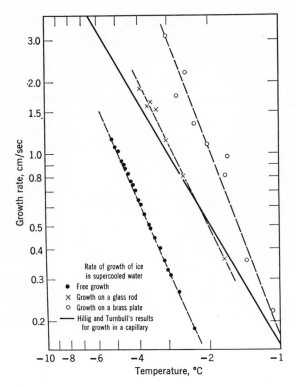

Fig. 4.16. Effect of substrate on rate of growth of ice in supercooled water. (From Ref. 22, p. 822.)

point P at approximately the same time. If the speed of growth of the secondary arm were equal to that of the primary, or $PB = PA$, then the envelope of the dendrite would be a regular hexagon without re-entrant angles as shown in Fig. 4.15c. The only difference between the primary spike OA and the secondary PB is that the latter is one of a parallel array, and it is concluded that each is retarded by the thermal field of its neighbors. The ratio of the speeds of the free spike (OA) and the retarded spike (OB) is given by the angle PAB. It is observed that this angle is less than 60° only at small undercooling, showing that the mutual interference of neighboring spikes is important only when growth is slow.

It will be evident that the growth velocity of a dendrite that is a member of a row may be less than that of a single isolated dendrite, and that each dendrite of a two-dimensional array will grow even more slowly as a result of their mutual interference by the overlapping of their thermal fields.

EXPERIMENTAL OBSERVATIONS OF RATE OF DENDRITIC GROWTH. Many observations have been made on the speed of growth of crystals in supercooled liquids; many of these experiments have given information on the growth of crystals in contact with a solid substrate, conditions which, as shown by Lindenmeyer et al. (22), bear little relationship to the results obtained in free growth in the liquid. Figure 4.16 shows the results of experiments on the rate of growth of ice in water both with and without substrates.

The only experimental observations on rates of growth of freely growing dendrites are the following:

(a) Weinberg and Chalmers (3) showed that the lead dendrites grew much faster than the smooth interface.

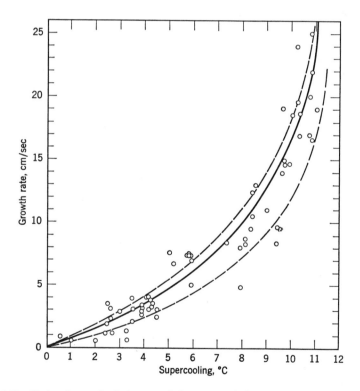

Fig. 4.17. Rate of growth of tin crystals in supercooled liquid tin. (From Ref. 23, p. 343.)

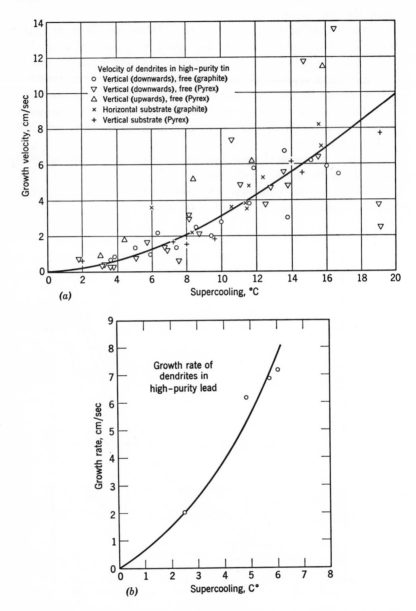

Fig. 4.18. Rate of growth of tin in supercooled liquid tin and lead. (From Ref. 24.)

(b) Rosenberg and Winegard (23), by observation of the distur-bance to the surface of a supercooled bath of tin by growing dendrites, measured the growth rate in the temperature range $\Delta T = 0.4°$ to $\Delta T = 11°$; their results are shown in Fig. 4.17.

It will be seen that there is substantial scatter, a factor of three exist-ing between the highest and the lowest values for a given temperature. Part of this scatter may, perhaps, be accounted for on the basis that the dendrites, which were nucleated by local cooling of the melt, grew with random orientations; since the dendrites in lead always grow in cube directions, the distance traveled by a dendrite while advancing one unit of distance in the $\langle 111 \rangle$ direction would be $\sqrt{3}$ units. The scatter is, however, greater than would be accounted for in this way, and it is probable that the measured temperature of the bulk liquid did not properly represent the actual temperature of the surface where growth was observed. It is also possible that growth at a free surface may differ from that in the liquid in an inverse way to that of growth on a solid substrate.

(c) Orrok's measurements on tin (24) were made within the bulk rather than at the surface; he timed the rise of temperature at two points in the melt, observed from the output of thermocouples con-nected to an oscilloscope circuit. His results are shown in Fig. 4.18.

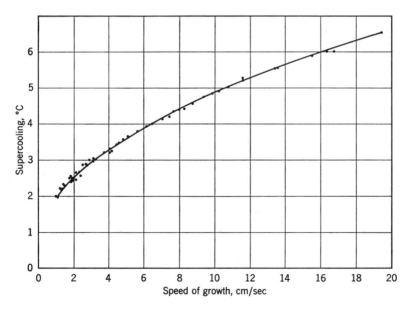

Fig. 4.19. Rate of growth of ice in supercooled water. (From Ref. 25.)

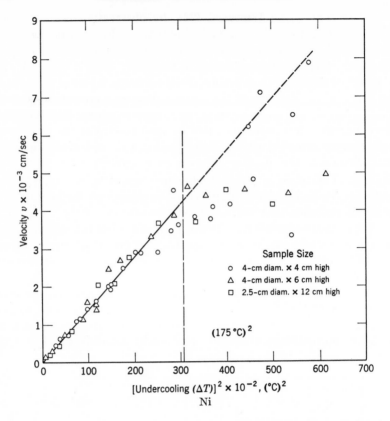

Fig. 4.20. Rate of growth of nickel in undercooled melt. (From Ref. 26.)

It will be noticed that there is again considerable scatter, and that the results differ considerably from those of Rosenberg and Winegard.

(*d*) Lindenmeyer (25) made extensive measurements on the growth of ice in supercooled water, using visual observation of the dendrites growing vertically downwards in a glass tube. Typical results are shown in Fig. 4.19.

(*e*) Walker (26) has made extensive measurements on the growth of dendrites in nickel and cobalt, both of which he has succeeded in cooling in bulk (400 gm) to the homogeneous nucleation temperature. Walker initiated solidification at the free surface of the melt and observed the time interval between recalescence at two points in the melt, using quartz rods to transmit the light to photocells, connected to an oscilloscope. Walker's results for nickel are shown in Fig. 4.20, in which the growth velocity is plotted against the square of the supercooling; it is found that a good linear relationship is obtained from zero to about 175° supercooling; beyond that point, the results fall into 2 classes. Some points fall on the $v \propto (\Delta T)^2$ line, while others fall below that line but above 4000 cm/sec. As indicated in Fig. 4.19, Walker's results were obtained with 3 different sizes of specimen, a useful confirmation that surface growth accelerated by heat transfer to the container is not influencing the result significantly. These results show less scatter and cover a larger range of super-cooling than any previous data, and their support for the $(\Delta T)^2$ law is clear, except for supercooling in excess of 175°, the significance of which is discussed on page 121. Walker's results for cobalt, shown in Fig. 4.21, show similar characteristics; the velocity is again linear with $(\Delta T)^2$.

(*f*) Colligan and Bayles (27) also carried out growth velocity measurements on nickel, both by a method using optical fibers to transmit the light due to recalescence to a photocell (similar to the method used by Walker), and by high speed photography of the top surface of the melt. The results obtained by both of these methods are shown in Fig. 4.22.

EXPERIMENTAL OBSERVATIONS ON THE RATE OF GROWTH OF RIBBON CRYSTALS OF GERMANIUM. It has been clearly demonstrated by Seidensticker and Hamilton (28, 29, 30) that the growth of germanium

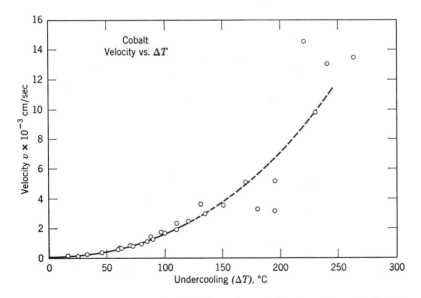

Fig. 4.21. Rate of growth of cobalt in undercooled melt. (From Ref. 26.)

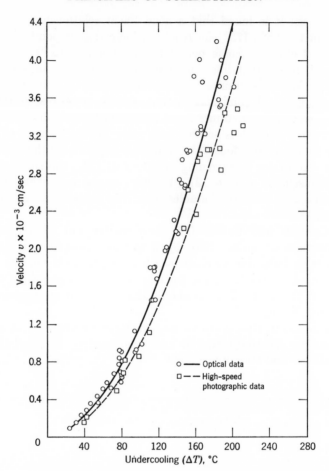

Fig. 4.22. Rate of growth of nickel in supercooled melt. (From Ref. 27, p. 896.)

"dendrites" of the ribbon type can be accounted for quantitatively on the steady state theory of growth of a paraboloid of revolution if the measured tip radius is used. This radius, however, does not coincide with the value predicted on the maximum velocity criterion if the kinetic driving force is neglected. It is concluded that although the departure of the tip temperature from equilibrium is small, this difference has an important influence on the tip radius and therefore on the rate of growth.

Direction of dendritic growth. It was pointed out by Weinberg and Chalmers (3) that the arms of dendrites always grow in crystallographically determined directions, each of which is the axis of a pyramid whose sides are the most closely packed planes with which a pyramid can be formed (this excludes the basal plane in the hexagonal structure) (31). The directions that conform to this description, and are in fact observed experimentally, are given in Table 4.1.

Table 4.1.   Direction of Dendritic Growth

Structure	Dendritic Growth
Face-centered cubic	$\langle 100 \rangle$
Body-centered cubic	$\langle 100 \rangle$
Hexagonal close-packed	$\langle 10\bar{1}0 \rangle$
Body-centered tetragonal (tin)	$\langle 110 \rangle$

According to these generalizations, the dendrite arms should always be orthogonal in the cubic and tetragonal structures and should form angles of 60° for the hexagonal close-packed metals. These rules are always fulfilled when the dendrites are exposed by decanting the liquid from a melt in which dendrites are growing; however, dendrites are often seen at a surface, either a free surface or one that was, during solidification, in contact with a mold wall. In such cases the arms

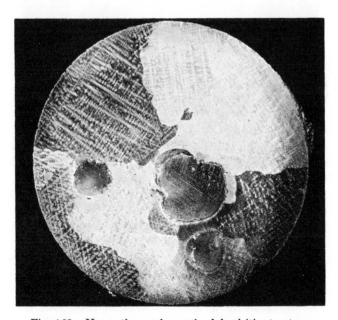

Fig. 4.23.   Non-orthogonal growth of dendritic structure.

Fig. 4.24. Independence of growth direction and thermal conditions.

are not always orthogonal because the structure seen at the surface represents the intersection with the surface of the three sets of cube planes, in the case of a cubic structure. An example of a "non-orthogonal" dendritic structure in an aluminum alloy is shown in Fig. 4.23. This structure can be explained on the basis that when a dendrite arm reaches a surface, and cannot grow farther, its branches in the two orthogonal directions grow instead; they branch parallel to the original arm, as a series of branches that meet the surface at its intersection with the appropriate {100} planes.

The general explanation for the crystallographic features of dendritic growth must be related to anisotropy of the relationship between growth rate and kinetic driving force. If there were no anisotropy, the dendrite should grow in a direction that is controlled entirely by thermal conditions; that this is not so is clearly seen in Fig. 4.24 which shows the dendritic structures in two crystals that grew side by side as a bicrystal into a supercooled melt of lead. It is evident that the characteristic directions are quite differently related to the heat flow pattern in the left and right halves of the specimen.

The observed habit in face-centered cubic crystals would be accounted for if it could be shown that the growth rate, for a given ΔT, is greater for $\langle 100 \rangle$ directions than in $\langle 111 \rangle$ directions. Then the forward growth in a $\langle 100 \rangle$ direction would be restrained by the slower growth of the $\{111\}$ planes that would form the sides of the pyramid truncated by a $\{100\}$ plane as shown in Fig. 4.24. The "idealized" form of the dendrite shown in Fig. 4.25a would not, of course, conform to the heat flow conditions required for steady state growth, but the "rounded off" form of Fig. 4.25b would do so if a correction could be made for the slower growth (for a given supercooling) of the $\{111\}$ faces than the $\{100\}$ faces. The transition between the $\{100\}$ and the $\{111\}$ faces is through high index or noncrystallographic surfaces that could, presumably, grow even faster than the $\{111\}$ face at the tip.

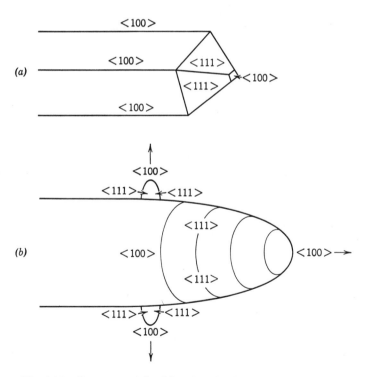

Fig. 4.25. Geometry of dendrite tip. (a) Schematic, (b) actual.

The existence of some anisotropy in growth rate between more and less closely packed faces is to be expected on the basis, proposed by Cahn (32) that at low driving force, the surface of a metal, while diffuse, still grows by the lateral propagation of steps. This allows a distinction to be retained between different faces; this would presumably cease to exist if the faces became completely diffuse and advanced entirely without step propagation. Bolling and Tiller (11) conclude that the driving force for dendritic growth in metals is small enough for the diffuse step mechanism to operate.

There is, so far, no way of measuring the anisotropy of growth rate that actually exists; neither is there any analysis that indicates how much anisotropy is required to account for the observed independence of growth direction on the thermal gradients that must always be present in the melt.

DIRECT OBSERVATION OF DENDRITE SHAPE. Several attempts have been made to determine the shapes of dendrites during growth by rapidly separating the dendrite from the liquid and then examining it microscopically. Atwater and Chalmers (33) observed surface markings on lead dendrites withdrawn from the melt and identified them as the traces of {111} and {100} planes; they concluded that the surfaces of the dendrites, as examined, were terraced, the terraces corresponding to the "cropping out" of all four sets of {111} planes and one set of {100} planes. They were not satisfied that these terraces were present as features of the interface during growth, because the terraces could have formed during the solidification of the liquid layer that adheres to the interface during separation from the melt. Bolling and Tiller (15) also observed a terraced structure on lead dendrites and identified the terraces as close-packed (i.e., {111}) planes. They concluded that the terraces were present during growth of the dendrite. However, it has been shown more recently by Chadwick (34) that there is no justification for identifying the "platelet" or "terraced" structure seen on the surface of decanted specimens as a structure that was actually present during growth.

Spacing of dendrite arms. There is very little experimental information on the spacing of dendrite arms in pure metals; the most direct is that of Weinberg and Chalmers (3) who showed that the spacing of primary dendrite arms in lead increases as the supercooling increases.

This increase in spacing can be explained qualitatively on the basis that the development of a local perturbation of the surface into the instability that becomes a dendrite arm requires a temperature inver-

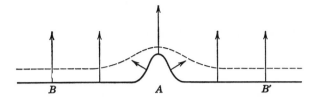

Fig. 4.26. Temperature distribution near a dendrite branch.

sion near the surface with a gradient that exceeds some finite minimum value. The region of the surface surrounding a developing instability will always have a smaller gradient of supercooling than regions that are undisturbed; this is illustrated in Fig. 4.26, in which the heat flow from a surface perturbed at A is shown, and an isotherm is indicated. It will be seen that the isotherm is disturbed from B to B', beyond which points the gradient of supercooling would have the value that was sufficient to develop the instability at A. It follows that no further arms can develop between B and B', and that the minimum spacing of branches would be AB. The effect of increasing the supercooling is to increase the rate of production of latent heat and therefore to increase distance AB in which the gradient is insufficient to allow arms to develop.

If the "instability" theory for dendritic growth is correct, then it would follow that the distance between successive positions of the tip at which instability developed would decrease as the supercooling increased, and it would follow that the process proposed for the production of branches would be more closely spaced, because the critical radius is smaller for larger supercooling. It is often observed, however, that some branches are suppressed by their neighbors, and it is likely that the *survival* of branches, rather than their *initiation*, is controlled by the thermal process described above. There is as yet no satisfactory quantitative theory of branch spacing.

Similar arguments would apply to the spacing of the branches that grow laterally from the main arms, and to the secondary and successive generations of branches. The experimental observation is that each successive generation is on a progressively finer scale, that is, thinner spikes more closely spaced. This would correspond to the expected progressive decrease in the supercooling of the remaining liquid as growth proceeds. The branches on any individual arms are approximately, but not precisely, uniformly spaced.

4.6   Solidification at Very High Supercooling

In his study of the solidification of supercooled nickel, described
on page 115, Walker (26) found that a new regime in the velocity-
supercooling relationship appears when the supercooling exceeds about
175°; a similar change in behavior occurs in cobalt at about the same
supercooling.   That the mode of solidification undergoes a profound
change is shown by the size of the crystals in the solidified metal.
Walker has shown that in the "low" supercooling region (0° to 175°)
there are only a few crystals in the whole specimen, a typical grain
diameter being 1.5 cm.   In the "high" supercooling range (>175°) the
grain size is much finer, typically 0.01 cm.   The variation of grain size
with supercooling is shown in Fig. 4.27.   It is interesting to note that
reliable estimates of the grain-size characteristic of the high supercooling conditions could not be obtained without "doping" the melt
with a small addition of silver, which inhibits grain growth after
solidification.

Colligan and Bayles (27) confirmed these results and obtained photo-
graphic evidence that the solid-liquid interface, as seen at the top
surface of the solidifying melt, consists of a few distinct dendrites
for the low supercooling region while at high supercooling it becomes
smooth and continuous, as would be expected for a fine grain size.

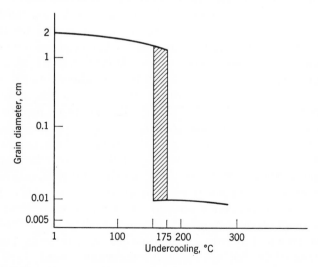

Fig. 4.27.   Grain sized nickel as a function of undercooling.   (From Ref. 26.)

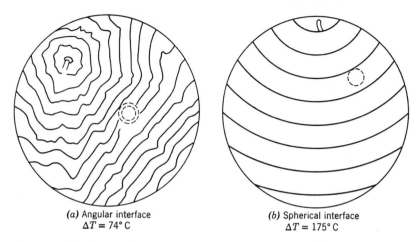

(a) Angular interface
$\Delta T = 74°$ C

(b) Spherical interface
$\Delta T = 175°$ C

Fig. 4.28. Successive positions of the interface of nickel solidifying from *(a)* small supercooling, *(b)* large supercooling. (From Ref. 27.)

Examples of the appearance of the two types of interface are shown in Fig. 4.28.

It may be concluded from the experimental results that when the supercooling (for nickel) is greater than 175°, the process of dendritic growth of existing crystals may be interrupted by the nucleation of new crystals, where growth is again interrupted by the formation of more crystals; on the other hand, Walker's results indicate that the "low" supercooling behavior may sometimes persist at substantially greater supercooling. Horvay (35) has proposed a theory to account for the nucleation of new crystals at the surface of a sufficiently rapidly growing crystal. He considers the ideal case of an undercooled incompressible liquid which increases in density on solidification, and he analyzes the pressure changes in the liquid that occur, as a result of the volume change, near a crystal growing in the melt. He takes the case of a crystal starting at zero size and growing with a planar, cylindrical, or spherical front at a rate, predicted from heat flow considerations, that is proportional to $t^{1/2}$, where t is the time since the beginning of growth. Horvay shows that a very large negative pressure can develop in the liquid adjacent to the growing crystal; this negative pressure can be large enough to cause cavitation, if the supercooling is sufficient; cavitation, as shown in Chapter 3, may cause nucleation. Horvay proposes that the condition for cavitation actually to occur is that the radius of the growing crystal should be above its critical nucleation radius when the cavitation pressure is reached;

and he shows that this should occur for nickel at about 175° super-cooling. The theory shows that as the crystal grows, the negative pressure passes through a maximum value. It is possible that the occasional very high values of velocity are the result of dendrites growing initially in a region of less supercooling than that of the melt as a whole, so that they do not encounter the very high supercooling until they are above the critical size for cavitation at that tempera-ture. They could grow at the rate controlled by heat flow, which re-sults in anomalous adherence to the $(\Delta T)^2$ law.

It is probably significant that although the assumptions of the theory in its present form are, as emphasized by Horvay, unrealistic, it does account for both the normal behavior at high supercooling and the occasional anomalously high velocities.

Some interesting acoustical observations by Walker give apparent support to the cavitation theory of limiting dendrite speeds; he ob-serves (26) that an audible click accompanies the solidification of nickel when there is substantial supercooling; a rough measure-ment of the intensity of the sound shows that it varies with super-

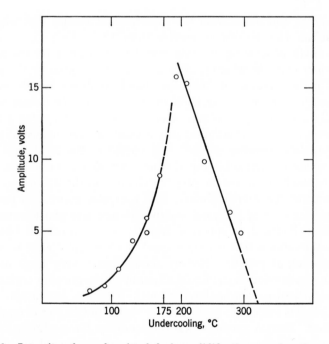

Fig. 4.29. Intensity of sound emitted during solidification as a function of super-cooling. (From Ref. 26.)

cooling as shown in Fig. 4.29. It is possible that the sound may be generated by the collapsing cavities; it is also possible that the sound is generated as the solid shrinks away from the mold wall.

References

1. For reference to this, and other historical aspects of dendritic solidification, see C. S. Smith, *History of Metallography:* University of Chicago Press, 1960.
2. F. Weinberg and B. Chalmers, *Can. J. Phys.,* **29,** 382 (1951).
3. F. Weinberg and B. Chalmers, *Can. J. Phys.,* **30,** 488 (1952).
4. J. O'Connor, *Electrochem. Soc.,* **107,** 713 (1961).
5. G. F. Bolling and W. A. Tiller, *J.A.P.,* **31,** 1345 (1960).
6. B. Chalmers, *Proc. Roy. Soc., Ser. A,* **175,** 100 (1940).
7. B. Chalmers, *Proc. Roy. Soc., Ser. A,* **196,** 64 (1949).
8. F. S. Ham, *J. Chem. Phys. Solids,* **6,** 335 (1958).
9. W. W. Mullins and R. F. Sekerka, *J.A.P.,* **34,** 323 (1963).
10. R. S. Wagner, *Acta Met.,* **8,** 57 (1960).
11. J. Herenguel, *Rev. Metallurgie,* **45,** 139 (1948).
12. J. Herenguel and P. Lacombe, *Comptes Rend.,* **228,** 846 (1949).
13. K. T. Aust, F. M. Krill, and F. R. Morral, *J. Met.,* **1952,** 865.
14. W. Roth and M. Schippers, *Z. Metall.,* **47,** 78 (1956).
15. K. A. Jackson, *Private Communication.*
16. J. C. Fisher, *Private Communication* (1950).
17. B. Chalmers and K. A. Jackson, *Private Communication* (1953).
18. G. Horvay and J. W. Cahn, *Acta Met.,* **9,** 695 (1961).
19. G. F. Bolling and W. A. Tiller, *J.A.P.,* **32,** 2587 (1961).
20. D. E. Timken, *Soviet Phys., Doklad.,* **5,** 609 (1960).
21. J. W. Cahn, *Private Communication.*
22. C. S. Lindenmeyer, G. T. Orrok, K. A. Jackson, and B. Chalmers, *J. Chem. Phys.,* **27,** 822 (1957).
23. A. Rosenberg and W. C. Winegard, *Acta Met.,* **2,** 342 (1954).
24. T. Orrok (Thesis). *Dendritic Solidification of Metals,* Harvard, 1958.
25. C. S. Lindenmeyer (Thesis). *The Solidification of Supercooled Aqueous Solutions,* Harvard, 1959.
26. J. L. Walker, *Private Communication.*
27. G. A. Colligan and B. S. Bayles, *Acta Met.,* **10,** 895 (1962).
28. R. G. Seidensticker and D. R. Hamilton, *J.A.P.,* **31,** 1165 (1960).
29. R. G. Seidensticker and D. R. Hamilton, *J.A.P.,* **34,** 1450 (1963).
30. R. G. Seidensticker and D. R. Hamilton, *J.A.P.,* **34,** 3113 (1963).
31. B. Chalmers, *Trans. AIME,* 519 (1954).
32. J. W. Cahn, *Acta Met.,* **8,** 554 (1960).
33. H. A. Atwater and B. Chalmers (unpublished), 1957.
34. G. Chadwick, *Acta Met.,* **10,** 1 (1962).
35. G. Horvay, *ASME Proc. 4th National Congress of Applied Mechanics,* 1962, p. 1315.

5

Redistribution
of Solute during
Solidification

5.1 General Considerations

Equilibrium between a crystalline solid and a liquid is conveniently represented on a binary equilibrium diagram by means of two lines, the liquidus line, above which the liquid is the stable phase, and the solidus line, below which the solid is stable. For ternary systems, the liquidus and solidus conditions are defined by surfaces; in general, as shown by the phase rule, the liquidus and solidus conditions in an N-component system are determined by $N - 1$ independently variable parameters. The fact that the liquidus and solidus do not coincide, except for pure materials and in the exceptional cases of congruent melting, indicates that a solid usually differs in composition from the liquid in equilibrium with it. A consequence is that when a liquid solution, initially of uniform composition, is solidified progressively, the composition of the solid is not uniform; the distribution of solute in the solid, when solidification is complete, is different from that in the liquid, although the total amount of solute is unchanged.

5.2 The Distribution Coefficient

It is convenient to start the discussion of the way in which the solute is redistributed by considering an idealized case in which the liquidus and the solidus lines are both straight, as illustrated in Fig. 5.1, where the two possible cases are shown. The type shown in Fig. 5.1a is much more common than that of Fig. 5.1b, but the discussion will relate to both cases. It is convenient to describe the salient feature of these solid-liquid equilibrium relationships by means of the distribution coefficient or partition coefficient, k, a term for which two distinct

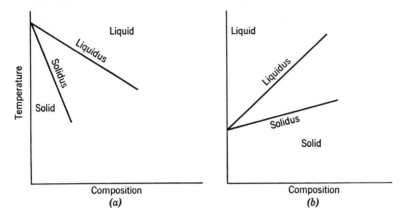

Fig. 5.1. Solidus-liquidus relationships for dilute binary alloys.

meanings have arisen. The definitions used here are as follows: the *equilibrium* distribution coefficient, k_0, is the ratio of concentrations of the solid and the liquid in equilibrium with it, i.e., C_S/C_L in Fig. 5.2; the *effective* distribution coefficient k_E, is given by

$$k_E = \frac{C_S}{C_0}$$

where C_S is the concentration of the solid that is formed at some instant by solidification of a liquid of average concentration C_0. It will be shown below that the value of k_E depends on the conditions under which solidification takes place, while k_0 is a characteristic of the system. k_0 is not necessarily a constant for a given system, because

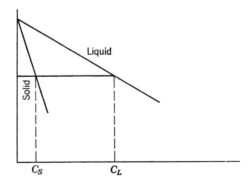

Fig. 5.2. Relationship between equilibrium compositions of solid and liquid.

the liquidus and solidus lines may not be so related as to maintain a constant ratio.

Although the quantity k_0 has been defined in terms of equilibrium, it has been shown by Jackson (1) that the ratio of concentrations of the solid and the liquid *in contact with the solid* is changed only very slightly when solidification or melting takes place; the change caused by motion of the interface at speeds that are normally attained experimentally is less than the probable error with which the value of k_0 has been measured. It will therefore be assumed that k_0 is independent of the speed of motion of the interface.

It has, however, been shown recently by Biloni and Chalmers (2) that when a liquid is sufficiently supercooled (probably below its solidus temperature) the first nuclei to form grow as crystals *with the composition of the liquid* until the temperature rises, as a result of the evolution of latent heat, to a level at which the equilibrium relationship comes into play. This process of "solidification without diffusion" is possible because the decrease in free energy that is the thermodynamic driving force for solidification is available without change of composition if the supercooling is sufficient, whereas the usual equilibrium condition implies that the decrease in free energy is conditional upon the change of composition corresponding to k_0. The process under discussion is analogous to the so-called "massive transformation" that has been identified in solid systems.

5.3 Rejection of Solute

It is a necessary consequence of the equilibrium relationship between solids and liquids, represented by the phase diagram (Fig. 5.3) that the composition.of the liquid must change when solid is formed from it;

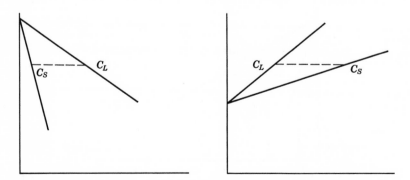

Fig. 5.3. Composition of solid formed from liquid of composition C_L.

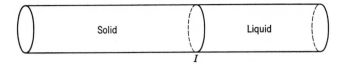

Fig. 5.4. Geometry of uniaxial solidification.

in the case illustrated in Fig. 5.3a, solute is rejected by the solid as it forms, while in Fig. 5.3b solvent is rejected at the solid-liquid interface. Various assumptions can be made regarding the subsequent motion, in the liquid, of the solute or solvent that is rejected at the interface; the extreme assumptions are (a) that it is at all times completely mixed with the whole of the remaining liquid, and (b) that transport is by diffusion only. Assumptions can also be made regarding the motion of solute, by diffusion, in the solid. Each of the possible cases will be examined for the geometrically very simple system represented by Fig. 5.4, in which the interface is assumed to be planar and to move at constant speed, and in which the composition is assumed to be uniform across any cross section (parallel to the interface). This is equivalent to assuming that the transport of solute (or solvent) is parallel to the direction of motion of the interface; that is, that both heat flow and solute diffusion are uniaxial.

Equilibrium maintained at all times. When solidification starts (it is assumed that there is no nucleation problem) the composition of the liquid is C_L (Fig. 5.5) and the first solid that forms, in equilibrium

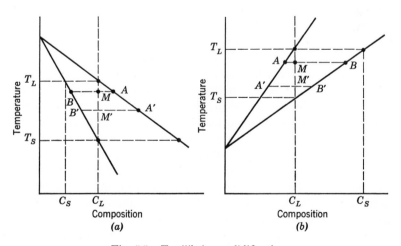

Fig. 5.5. Equilibrium solidification.

with it, has a composition C_S. The formation of a finite amount of solid (of composition C_S) must be accompanied by a change in the composition of the liquid; the concentration of solute in the liquid increases and decreases in the cases represented by Fig. 5.5a and Fig. 5.5b, respectively. The direction of the change is always such that the liquidus temperature falls.

The case under consideration is defined by the assumption that the interface advances so slowly that the rejected solute is uniformly mixed into the whole of the liquid at all times, by fluid motion and diffusion, and that the diffusion of solute in the solid is sufficient to keep the whole of the solid at the composition that is in equilibrium with the liquid. Thus when the composition of the liquid has moved to the point A (Fig. 5.5) that of the whole of the solid will have reached the point B. As solidification progresses, the points A and B move down their respective lines, to positions such as $A'B'$. Since the total amounts of solvent and solute remain unchanged, the relative amounts of solid and liquid are always given by the ratio AM/BM; and it follows that the process is complete (i.e., no liquid remains) when the point B reaches the line MC_L, that is, when the whole of the solid has the composition that the liquid had originally. The temperature at which the process finishes is the solidus temperature for the original composition and the resulting solid is of uniform composition, identical to that of the original liquid.

The assumption that the whole of the solid is at all times in equilibrium with the liquid implies that there is no concentration gradient in the solid, although the process of solidification itself tends to generate one by depositing solid from a liquid with a continuously changing composition. The assumption, therefore, demands that the process of diffusion in the solid is fast enough to reduce the concentration gradient to a negligibly small value. This can happen only if the rate of advance of the interface is slow compared with the diffusion rate of the relevant solute in the solid, and if the required diffusion distance (total distance through which the interface moves) is small.

These conditions are never completely satisfied; the nearest approach is probably in geological processes, in which the time scale is extremely long by human or industrial standards. Among metallurgical processes, the largest effect of diffusion during solidification is probably to be found in the case of carbon and nitrogen in steel, since these interstitial solutes have very much higher diffusion coefficients than substitutional solutes. It should also be remembered that a significant amount of diffusion can also take place after solidification while the material is cooling down. This is discussed on page 178.

Mixing in the liquid by diffusion only; no diffusion in the solid. The opposite extreme to the previous case would be that in which the solute (or solvent) does not move, and there is consequently no mixing in either the solid or the liquid. This would imply that the liquid changes into solid without change of composition. It has been shown that "diffusionless solidification" of this kind may take place locally when the liquid is sufficiently supercooled. However, in the geometrically simple system under discussion, the nearest possible approach to solidification without mixing is that in which transport is by diffusion only. Since solid diffusion is usually much slower than liquid diffusion, the former will be ignored, and it will be assumed that all diffusion in the liquid is normal to the interface.

It is necessary, in order to consider this case, to introduce the speed of advance of the interface, R cm/sec. This variable was not relevant to the previous case because it was implicitly assumed to be vanishingly small. The distribution of solute in the liquid at any time takes the form shown in Fig. 5.6a, in which the compositions of the solid and the liquid at the interface are C_s and C_a, respectively. The ratio C_s/C_a is equal to k_0. The initial composition of the liquid is C_0, and k_E is equal to C_s/C_0. The solute rejected at the interface diffuses into the liquid, and its distribution is represented by the curve D.

It will be evident that if $C_s = C_0$, as in Fig. 5.6b, the amount of solute taking part in the diffusion process is constant. That is, it does not change as the interface moves to the right. This amount is represented by the shaded region of Fig. 5.6. Thus it follows that a steady state condition can exist, at which solid of composition C_0 is formed.

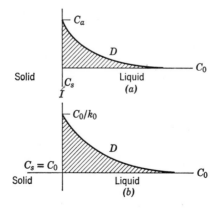

Fig. 5.6. Distribution of solute during uniaxial solidification.

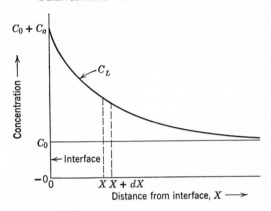

Fig. 5.7. Steady state diffusion of solute. (From Ref. 3.)

It is convenient first to consider the steady state conditions in detail, and then to examine the various transient processes.

STEADY STATE CONDITIONS. Under steady state conditions, the amount of solute rejected at the interface is just balanced by the amount that diffuses away from it; the diffusion coefficient for the solute in the liquid is D cm²/sec.

The steady state distribution of solute ahead of the interface can be found as follows (3). Let the curve of Fig. 5.7 represent the steady state distribution of solute; the amount of solute diffusing into the unit area of face X is $D(dC/dx)_x$, and the amount diffusing out is $D(dC/dx)_{x+dx}$; there is, therefore, a net flow of $D(d^2C/dx^2)$ per unit volume into a volume element. If the solid-liquid interface is taken as the origin, and solidification is represented by moving the solute distribution past any point at the rate of R cm/sec, then the net flow out of the same element is $R(dC/dx)$. For steady state, the sum of the amount flowing in and the amount flowing out must be zero, hence

$$D \frac{d^2C}{dx^2} + R \frac{dC}{dx} = 0$$

The solution to this equation is

$$C_L = C_a \exp\left(-\frac{R}{D} X'\right) + C_0$$

where C_a is the solute concentration in the liquid at the interface, and X' is the distance from the interface at which the concentration is C_L.

An essential feature of the steady state condition is that the solid is formed with the composition C_0, since only in this case does the amount of solute in the enriched layer remain constant. If the composition of the solid is C_0, then C_a must be C_0/k_0, and the expression for C_L becomes

$$C_L = C_0 \left[1 + \frac{1 - k_0}{k_0} \exp\left(-\frac{R}{D} X' \right) \right]$$

This relationship shows that the liquid distribution is exponential, with a "characteristic distance" given by D/R; that is, the distance in which the excess concentration falls to $1/e$ of its initial value.

Existing data indicate (4, 5) that the diffusion coefficients are the same within a factor of ten for all liquid metals; a representative value is 5×10^{-5} cm²/sec, or 5 cm²/day. Figure 5.8 shows how the characteristic distance varies with R, using the value $D = 5$ cm²/day. It will be seen that, for the speeds of solidification considered, the thickness of the layer in which there is a substantial change in composition varies from less than 0.1 mm to about 1 mm. Good qualitative experimental confirmation of the existence of the enriched layer in contact with an advancing solid-liquid interface has been obtained by Kohn and Philibert (6), who used an electron microprobe to examine the distribution to copper in an aluminum copper alloy quenched during solidification. A typical result is shown in Fig. 5.9. Quantitative agreement could not be established because the interface in

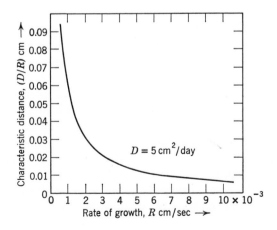

Fig. 5.8. Thickness of diffusion zone as a function of temperature. (From Ref. 3.)

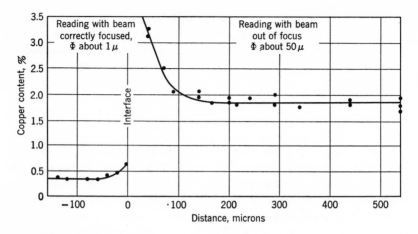

Fig. 5.9. Distribution of copper in an aluminum copper alloy quenched during solidification. (From Ref. 6.)

this case was not planar and the assumption of linear solute transport was therefore not satisfied.

It follows from the fact that, under steady state conditions, the solid forms with a composition C_0, the temperature of the interface during this process must be the *solidus* temperature for the original liquid (i.e., T_s of Fig. 5.5), with a small correction for the slight super-cooling required to "drive" the solidification process (see page 42).

INITIAL TRANSIENT. The first solid to form from a liquid of composition C_0 must have a composition $k_0 C_0$. It follows that the steady state condition, in which the concentrations of liquid and solid at the interface are C_0/k_0, and C_0, will not be reached until the solidification has proceeded far enough for a substantial amount of solute to have been rejected.

Figure 5.10 represents the initial conditions, two intermediate stages, and the steady state for an alloy with a value of $k_0 = 0.33$. The necessity for solute to be conserved requires that the two shaded areas be equal, for the shaded area in the solid region represents the deficit of solute in the solid, and the shaded area in the liquid corresponds to the excess in the liquid, compared with C_0. It was suggested by Tiller et al. (3) that the composition of the solid C_S be given as a function of the distance from the start of solidification by

$$C_S = C_0 \left\{ (1 - k_0) \left[1 - \exp\left(-k_0 \frac{R}{D} X \right) \right] + k_0 \right\}$$

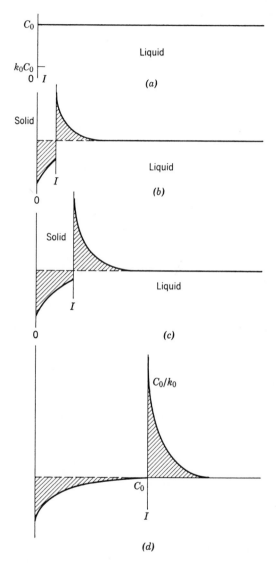

Fig. 5.10.   Distribution of solute during initial transient.

on the assumption that the relationship is exponential. It was later shown by Smith, Tiller, and Rutter (7) that the assumption of exponential approach to the steady state is not quite correct, and that an exact solution is

$$\frac{C_S}{C_0} = \frac{1}{2}\left\{1 + \text{erf }\sqrt{(R/2D)X} + (2k_0 - 1)\exp\left[-k_0(1 - k_0)\frac{R}{D}X\right]\right.$$
$$\left.\text{erf}\left[\frac{(2k_0 - 1)\sqrt{(R/D)X}}{2}\right]\right\}$$

The differences between the values for C_S calculated by these two methods are, however, so small that they can be ignored for almost all purposes, and the only significant aspect of the initial transient, that is, its characteristic distance X_c, can be found without serious error from

$$X_c = \frac{D}{k_0 R}\text{ cm}$$

TERMINAL TRANSIENT. Once the steady state has been achieved, it will be maintained so long as (a) there is sufficient liquid ahead of the interface for the forward diffusion of the solute to occur without hindrance, and (b) R remains constant. The former condition ceases to be satisfied when the boundary of the liquid is approached, as in Fig. 5.11, and it is evident that the concentration of the solid that is formed begins to rise above C_0 in order to accommodate the excess solute, which must all appear in the terminal region. Since the characteristic distance for the diffusion zone is D/R, and that for the initial transient is D/k_0R, it follows that the terminal transient zone occupies a shorter distance, by a factor k_0, than the initial zone. It can be

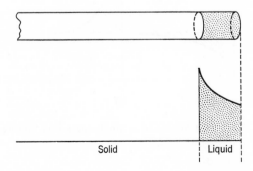

Solid   Liquid

Fig. 5.11. Distribution of solute during terminal transient.

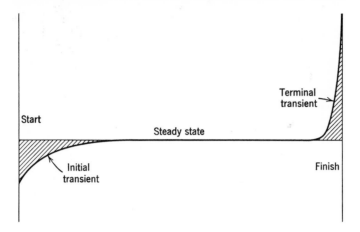

Fig. 5.12.   Concentrations in initial and terminal transients.

shown (qualitatively in Ref. 3 and in more detail in Ref. 7) that the concentration reached at the end of the terminal zone approaches the limit imposed by the alloy system as the quantity of liquid approaches zero; this is shown in Fig. 5.12.

There are no measurements of the change of concentration in the terminal transient, but there are numerous observations that can be explained readily in terms of a very high concentration of solute in the last liquid to solidify.   This question will be discussed further in Chapter 8.

CHANGE OF SPEED.   When R is constant, the amount of solute carried forward ahead of the interface, i.e., the area under the diffusion curve in Fig. 5.7, is proportional to the characteristic distance D/R, and is therefore inversely proportional to the rate of solidification R.

The solid that is formed immediately after a change of speed must, therefore, have a concentration that differs from C_0.   If the value of R is increased, D/R and the amount of solute carried forward decrease; the solid must have a higher concentration during the transition from steady state at lower R to steady state at higher R.   A decrease in R, conversely, generates a region in which the concentration is less than C_0, as shown in Fig. 5.13.   More detailed discussions of the extent and profile of the change in concentration resulting from changes in speed is given by Tiller et al. (2) and by Smith (6).

EFFECT OF CURVATURE OF THE INTERFACE.   The conclusion reached above, that the steady state conditions are $C_s = C_0$ and $T_I = T_s$ is valid only if the interface is planar.   If it is convex (center of curva-

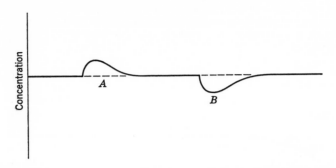

Fig. 5.13. Solute distribution following changes in speed of solidification. (A) Increase; (B) decrease.

ture in the solid), then the solute is not conserved within a cylindrical volume of the solidifying liquid (see Fig. 5.14). It follows that the composition of the solid that is being formed is below C_0, and that the concentration of the liquid adjacent to the interface is less than C_0/k_0. It also follows that the temperature of the interface is above the solidus temperature for the original liquid. Conversely, it follows that a concavity will have a higher solute content than C_0, and will solidify at a temperature below that of the initial solidus.

Influence of fluid motion: convection. The assumption of mixing by diffusion only is very unlikely to be realized in any practical case, because a liquid in which temperature gradients exist is likely to be subject to convection. The only case in which there would be no convection at all is that in which the density gradient in the liquid is everywhere vertical. If, in addition to density gradients caused by differences in temperature, there are also density gradients resulting from compositional variation caused by rejection of solute, the problem

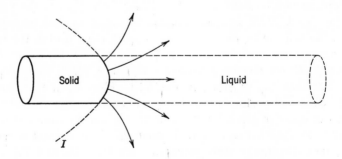

Fig. 5.14. Solute diffusion ahead of a convex interface.

of achieving a completely nonconvecting system becomes even more difficult. It is therefore necessary to consider the extent to which convection, resulting from density gradients that arise from nonuniformity of temperature or of composition, will affect the redistribution of solute during solidification.

The effect of mixing by fluid motion on the problem under discussion has been examined by Wagner (8), who assumed that the solute moves purely by diffusion through a layer of liquid of thickness d, beyond which there is sufficient convection to insure uniformity. The diffusion layer is the region, close to the interface, which is considered, in terms of hydrodynamics, to be "stagnant"; although it is not realistic to regard the liquid as stationary up to a distance d and in motion beyond it, the approximation is a useful one. Wagner shows that the thickness of the stagnant layer is sufficient to include nearly the whole of the diffusion zone, when the motion of the liquid is due to convection; in such cases, the analysis that ignores fluid motion is valid. When the motion of the liquid is more violent or when R is very small, however, the stagnant layer is not thick enough to accommodate the whole of the diffusion zone. The conditions may then be represented as in Fig. 5.15; diffusion limits the motion of solute between the interface and the point T, beyond which the liquid is mixed and has a composition C_p, which increases as solidification proceeds.

Complete or partial mixing of liquid: no diffusion in solid. A convenient way of handling the important intermediate case of the complete mixing of the whole of the liquid, except that in the stagnation zone, is to regard the liquid as being completely mixed at all times, and to allow for the existence of the diffusion layer by using an *effective distribution coefficient k_E*, instead of k_0 as in the previous

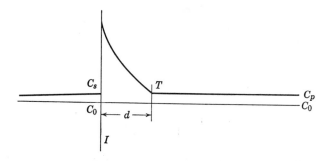

Fig. 5.15. Effect of stirring on the diffusion zone.

cases; k_E is defined as the ratio of the concentrations of solute in the solid that is formed to the average value for the liquid, which is approximated by that of the liquid remote from the interface. It can be shown (9) that the concentration C_S of the solid is given by

$$C_S = k_E C_0 (1 - g)^{k_0 - 1}$$

where g is the fraction of the liquid that has solidified in the system represented in Fig. 5.4. The variation of C_S/C_0 with g is shown in Fig. 5.16 for values of k_E from 0.01 to 5. It will be seen that there is no steady state region of the curve, a fact that is readily understood

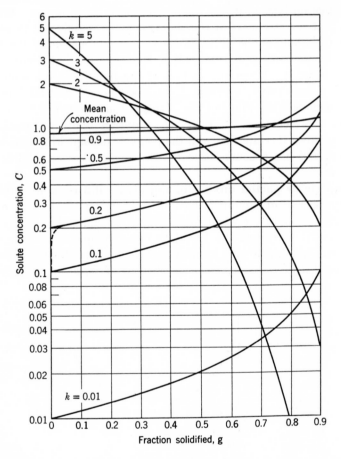

Fig. 5.16. Curves for normal freezing (liquid completely mixed at all times). C_0 is 1 for all curves. (From Ref. 9, p. 11.)

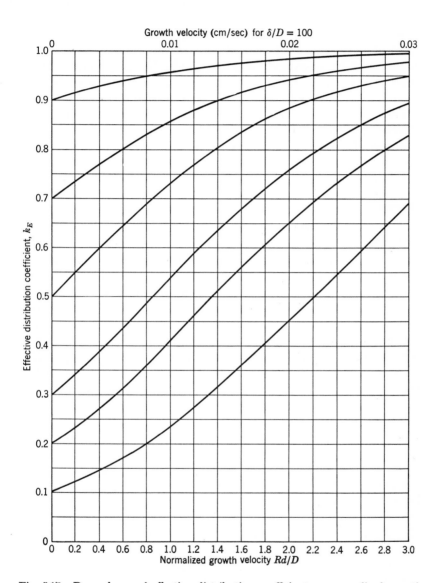

Fig. 5.17. Dependence of effective distribution coefficient on normalized growth velocity. (From Ref. 9, p. 14.)

on the basis that the composition of the whole of the liquid is changed continuously as rejected solute is mixed into it. It should also be noted that the speed of solidification does not appear explicitly in these curves; it does, however, influence the relationship because the value of k_E depends on the speed. At very low speeds, $k_E = k_0$; but as the speed increases and the diffusion zone becomes relatively more important, k_E approaches 1, the value it has during the steady state region of the "diffusion controlled" case discussed above.

The problem of calculating the value of k_E has been effectively attacked by Burton, Primm, and Slichter (10), who defined d as the thickness of the boundary layer through which the solute must diffuse; the value of d is limited by the velocity of the liquid parallel to the interface, and depends also on the viscosity of the liquid. It varies from about 10^{-3} cm, for very vigorous stirring, to about 10^{-1} cm for natural convection.

The value of k_E can be anywhere between k_0 and unity. It would be k_0 if the process were so slow or the mixing so effective that the whole of the liquid had the same composition as that in contact with the solid at the interface; on the other hand, if the bulk liquid remained at C_0, that is, not subject to any mixing, then the value of k_E would be unity. The value of k_E approaches k_0 as the effectiveness of mixing is increased; that is, as R is decreased, d is decreased, and the diffusion coefficient D is increased. It is therefore convenient to combine R, d, and D into the single dimensionless parameter Rd/D, which is used as a normalized growth velocity, instead of the actual growth velocity R, in the calculations of Burton, Primm, and Slichter, who obtained the expression

$$k_E = \frac{k_0}{k_0 + (1 - k_0) \exp\left(-\dfrac{Rd}{D}\right)}$$

Figure 5.17 shows how k_E varies with the normalized growth velocity for values of k_0 from 0.1 to 0.9. It will be seen that the treatment of the redistribution problem given here includes the extreme cases of complete mixing ($k_E = k_0$) and that of mixing by diffusion only ($k_E = 1$), as well as all intermediate cases, subject only to the conditions of uniformity of speed and uniaxiality of solute flow. The various cases are compared in Fig. 5.18.

Solidus temperature of an alloy. Many published phase diagrams include liquidus and solidus lines that were determined from cooling curves. While this method is satisfactory for the liquidus temperature, its use for solidus determinations can lead to large errors. The

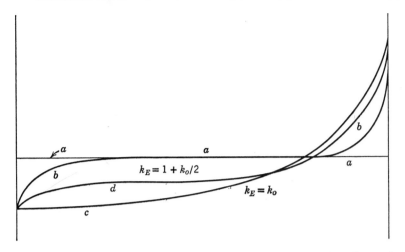

Fig. 5.18. Solute distribution after uniaxial solidification. (a) Complete diffusion in solid and liquid, (b) mixing by diffusion only, (c) complete mixing in liquid ($k_E = k_0$), (d) partial mixing in liquid [$k_E = (1 + k_0)/2$].

assumption underlying this method is that solidification begins at the liquidus temperature, or returns to it after some supercooling, and finishes at the solidus temperature. It will be clear from the foregoing discussion that when a single-phase solid is formed, the last liquid always solidifies at a temperature below the solidus for the original liquid (of composition C_0) sometimes by a large amount.

5.4 Zone Refining

The fact that a crystal growing from a solution usually rejects either the solute or the solvent has been used for very many years as a means of purifying crystalline materials. When a salt is allowed to

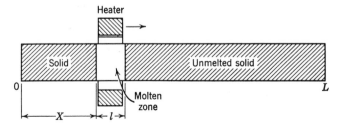

Fig. 5.19. Zone melting, schematic. (From Ref. 9, p. 24.)

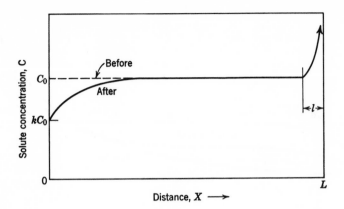

Fig. 5.20. Distribution of solute after passage of one molten zone; initial concentration C_0. (From Ref. 9, p. 25.)

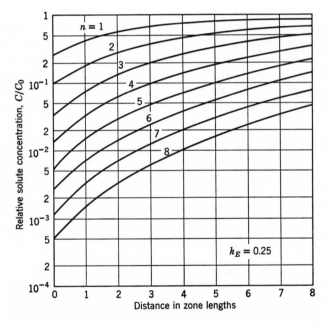

Fig. 5.21. Solute concentration against distance (in zone lengths) after passage of n zones ($k_E = 0.25$). (From Ref. 9, p. 30.)

crystallize from an aqueous solution, the value of the distribution coefficient for practically any solute is very small, and if islands of trapped solution can be avoided, a very satisfactory degree of purification can be achieved.   In cases such as these, the distribution coefficient is probably below 10^{-6}, and a single stage process is often adequate, but a second stage is sometimes used to reach even higher purity.   It would be possible to describe these processes in the terms used in the preceding pages, but because k_0 is so small, the values of R, d, and D are relatively unimportant.

In systems in which the distribution coefficient is substantially closer to unity, it is necessary to use a multi-stage process to achieve

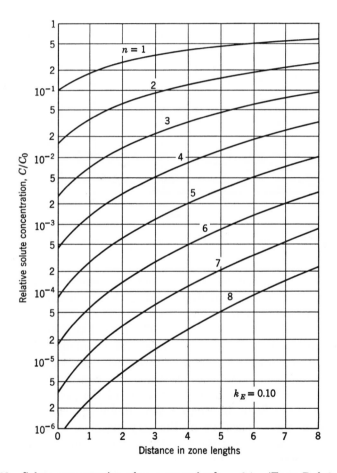

Fig. 5.22.   Solute concentration after n zones for $k_E = 0.1$.   (From Ref. 9, p. 31.)

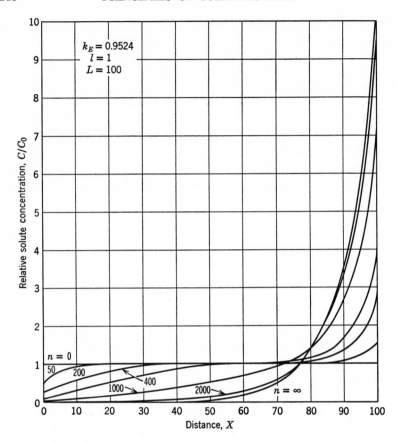

Fig. 5.23. Solute concentration for $k_E = 0.95$. (From Ref. 9, p. 34.)

high purity. It would be possible, but not practical, to purify an alloy by subjecting it repeatedly to unidirectional solidification, always rejecting the last part to solidify. Each successive stage would provide *less* material of *higher* purity. The part that is retained would be substantially purer at one end (the first to solidify) than at the other, and some of the separation of impurities that had been achieved during solidification would be lost when the retained portion is remelted as a preliminary to the next stage of the process.

The process of Zone Melting or Zone Refining invented by W. G. Pfann avoids these difficulties by the very ingenious device of moving a short molten zone along the bar instead of melting the whole of it. A detailed description of the process of zone refining, and its various extensions and modifications, is to be found in Pfann's book (9); an account of the basic principles will be given here. The metal is assumed to be in the form of a bar, held in a suitable container; it is heated by means of a ring heater so that a zone of length l is melted (Fig. 5.19). The result of passing a molten zone along the bar from left to right is to change the distribution of solute from a uniform value of C_0 to that shown in Fig. 5.20. The distribution is described by

$$\frac{C_s}{C_0} = 1 - (1 - k_E) \exp\left(-\frac{k_E X}{l}\right)$$

except in the terminal transient region, where a rapid increase in concentration takes place. The net result of the process is to transfer solute from near the "starting" end of the bar toward the "finishing" end.

If a second zone is passed along the bar in the same direction as the first, it causes further transfer of solute from left to right, and this continues with successive zones until the concentration gradients are everywhere such that the material formed by solidification of the molten zone has the same composition as the solid that goes into it. The distribution of solute after each of 8 successive passes of a zone, for a case of $k_E = 0.25$, is shown in Fig. 5.21. The distance along the bar is measured in multiples of the zone length. The effect of the value of k_E on the degree of purification is shown by comparing Fig. 5.22 and Fig. 5.23 with Fig. 5.21. It can readily be seen that for a

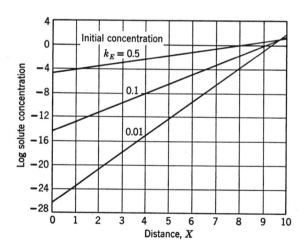

Fig. 5.24. Limiting distribution which is approached after passage of many zones; length of ingot is 10 zone lengths. (From Ref. 9, p. 41.)

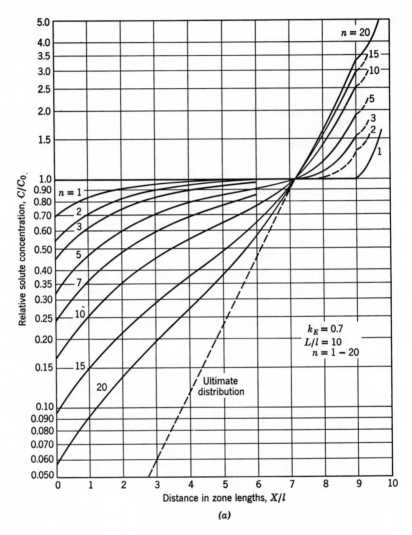

Fig. 5.25. Relative concentration after passage of n zones; (a) $k_E = 0.7$, (b) $k_E = 1.2$. [Part (a) from Ref. 9, p. 216; part (b) from Ref. 9, p. 220.]

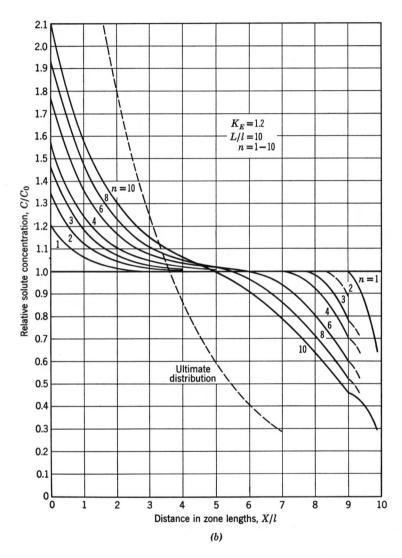

Fig. 5.25. (*Continued*)

low value of k_E (0.10) the concentration of solute is reduced by a large factor in a small number of passes, while a solute with a distribution coefficient close to 1, i.e., 0.95, requires a very large number to produce a comparable effect. Figure 5.24 shows how the ultimate distribution (i.e., when a steady state has been reached) varies with the value of k_E for a bar 10 zones in length. It has been assumed in the above discussion that the impurities to be removed have distribution coefficients less than 1; such solutes are rejected by the growing crystals and accumulate in the terminal region. There are, however, some solutes that are preferentially accepted by the growing crystals, and therefore have distribution coefficients greater than 1; these tend to accumulate in the first part to solidify. The value of k_E is seldom greater than about 2, at which value the separation would be about equivalent to that obtained at $k_E = 0.5$. A comparison of the separations obtained with $k_E = 1.5$ and $k_E = 0.7$ is shown in Fig. 5.25 in which a bar ten zones in length is subjected to various numbers of passes.

The very effective separation of solutes that can be obtained by zone refining makes it necessary to pay particular attention to the problem of avoiding contamination from the container in which the metal is melted. This problem becomes particularly acute with metals that are highly reactive and have high melting points. These difficulties can be overcome by using the various "floating zone" techniques that have been developed. In these methods, the molten zone is supported between the 2 solid parts of the bar, either by surface tension alone or by a combination of surface tension and forces of electromagnetic origin. For details, the reader is referred to Pfann's book and to the many detailed publications that have appeared.

5.5   Constitutional Supercooling

It has been shown that the liquid in contact with an advancing solid-liquid interface will in general have a composition that differs from that of the bulk liquid. Whether the distribution coefficient of a solute is greater or less than one, the liquidus temperature of the liquid in contact with the interface is lower than that of the liquid at a greater distance from the interface. It was pointed out by Rutter and Chalmers (11) that a consequence of this conclusion is that the temperature of the interface, in an alloy or an impure liquid, is lower than the liquidus temperature of the bulk liquid, and that supercooling can occur even if the temperature of the liquid is everywhere above that of the interface, i.e., if solidification is accompanied

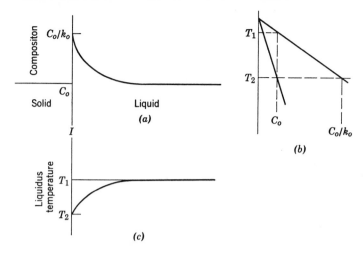

Fig. 5.26. Variation of concentration and liquidus temperature ahead of an interface. (a) Variation of concentration, (b) relationship between concentration and liquidus temperature, (c) variation of liquidus temperature.

by heat flow from the liquid to the solid. The concentration near the interface is shown in Fig. 5.26a, in which, for convenience, the steady state condition without convection has been selected. Figure 5.26b shows how the liquidus temperature varies with composition; i.e., it is a typical portion of a phase diagram for a binary alloy for which $k_0 = 0.2$. T_1 and T_2 are the liquidus temperatures for a composition C_0, and C_0/k_0; the former represents the bulk liquid, and the latter the liquid in contact with the interface. Figure 5.26c shows how the liquidus temperature varies with distance from the interface; the actual temperature of the interface itself is very close to T_2, differing from it only by the amount of supercooling necessary to provide the kinetic driving force for solidification; this difference will seldom amount to more than one-hundredth of one degree. The actual temperature distribution in the solid and the liquid may take the form shown in Fig. 5.27; this corresponds to a case in which the whole of the liquid is at a temperature above that of the interface; and yet *the liquid is supercooled* in the sense that it is below its liquidus temperature. Supercooling arising in this way is called "constitutional supercooling" (11, 12), to distinguish its origin from that of the type of supercooling discussed in Chapter 4, in which the liquid is at a lower temperature than the solid-liquid interface.

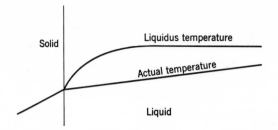

Fig. 5.27. Actual temperature of the liquid and its liquidus temperature.

Instability due to constitutional supercooling. It follows from the preceding discussion that, under the steady state conditions assumed, the solid-liquid interface will be very close to the *solidus* temperature of the alloy. If the interface remained planar, therefore, it should be possible to supercool the liquid by an amount equal to its "freezing range" of the alloy, that is, the interval between liquidus and solidus, while maintaining a positive temperature gradient in the liquid. The freezing range can be very large; for example, part of the equilibrium diagram for the copper-tin system is shown in Fig. 5.28, from which it is seen that an alloy of 90% copper, 10% tin has a freezing range of about 190°C. This would imply that the liquid ahead of the advancing interface could be constitutionally supercooled

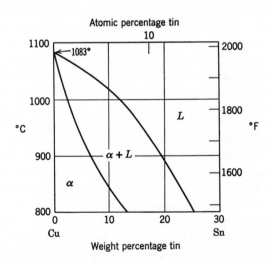

Fig. 5.28. Freezing range in copper-tin alloys. (From *Metals Handbook*, American Society for Metals, Cleveland, 1948, p. 1204.)

by 190°C. This condition is never even approached because a very much smaller amount of supercooling is sufficient to set up an instability which leads to a departure from the conditions of steady state advance of a planar interface which are implicit in the estimate of possible supercooling given above. The instability that develops resembles that discussed in Section 4.4 (p. 99) insofar as it may be described as a positive gradient of supercooling from the interface into the liquid.

Consequently, it is not to be expected that the solid which is formed will be of uniform composition; the solute content will instead vary in some periodic fashion. Two distinct types of periodicity have been discussed in the literature. We will first discuss the proposal, due to Landau (13) that the periodicity is axial; that is, that the solute content of the solid varies periodically in the direction of travel of the solid-liquid interface, and that it is uniform in any plane parallel to the interface. Landau assumes that the whole of the interface advances together, and shows that the accumulation of solute, as a result of the "initial transient," discussed on page 134, would increase progressively until some concentration is reached at which either independent nucleation would occur in the most supercooled region, or, in less extreme cases, the existing crystal would grow into that region. This would, in either case, produce a layer of high solute concentration as a result of fast advance of the interface. The liquid ahead of this would be much less enriched in solute, and so a new initial transient would occur. Thus there would be an alternation between rapid advance of the interface, with a resulting high concentration of solute, and slower advance accompanied by more effective forward diffusion of the solute. In support of this theory, Landau cites evidence for periodic variation of concentration of antimony in germanium crystals grown by the Kyropolous method (14), but no quantitative comparison between this theory and experiment is available. Many other examples exist in the literature of axially periodic variations of the solute content (15), but there is no conclusive evidence that these effects are in fact results of the Landau type of instability and not of periodic fluctuations in the thermal characteristics of the system, such as might be caused by the intermittent operation of a temperature control system. Albon (16), for example, has shown that axial segregation occurs in crystals of InSb grown by the "pulling" technique as a result of temperature fluctuations as small as 0.6 degrees; it is suggested that fluctuations of sufficient severity to cause "banding" could arise from the rotation of the crystal during pulling; Gatos, Strauss, Levine, and Harmon (17), however, have observed somewhat similar "banding" in unrotated samples. It is therefore concluded that there is some doubt as to whether the observed axial periodicity of composition is ever caused by the instability due to constitutional supercooling. If it does occur spontaneously, it is most likely to do so in materials in which the growth process is by lateral propagation of steps; i.e., smooth interface, because this type of growth tends to stabilize a flat or almost flat surface. This would oppose the alternative process which depends on the transverse (as distinct from axial) transport of solute.

5.6 Cellular Substructure

The alternative possibility is that the instability due to constitutional supercooling can be resolved by the development of a *transverse* periodicity in the solidification process; this concept has led to the explanation of a phenomenon that had been observed previously but had not been adequately accounted for. This is the development of a cellular substructure, first described by Smialowski (18) and also observed by Graf (19), Chalmers and Cahn (20), Goss and Weintraub (21), Pond and Kessler (22), Prince (23), and Rosi (24). The phenomenon can be readily observed on the top (free) surface of a crystal of tin grown from the melt. It appears as a fine regular corrugated structure, in which the corrugations (Fig. 5.29a) are roughly parallel to the direction of growth of the crystal. If the liquid is rapidly decanted, exposing the solid-liquid interface, the cellular structure shown in Fig. 5.29b is seen. It can readily be shown, by examining the upper edge of a decanted interface, that there is one to one correspondence between cells and corrugations.

It has been shown by Biloni (25) that the cellular structure takes the simple form shown in Fig. 5.29b only under conditions well within the range in which cells are produced; other types of cellular structure are shown in Fig. 5.29c.

Origin of the cellular substructure. No satisfactory explanation of the origin of the cellular substructure was proposed until it was shown by Rutter and Chalmers (11) that it can be accounted for in terms of constitutional supercooling, which, in the case of a planar interface, would occur as a gradient of supercooling from the interface into the liquid. A planar interface, however, would be unstable when exposed to a gradient of supercooling, because the growth rate would be increased in any localized region that advanced ahead of the general interface. Such a region would grow as a protuberance until, as a result of the gradient of supercooling combined with its own

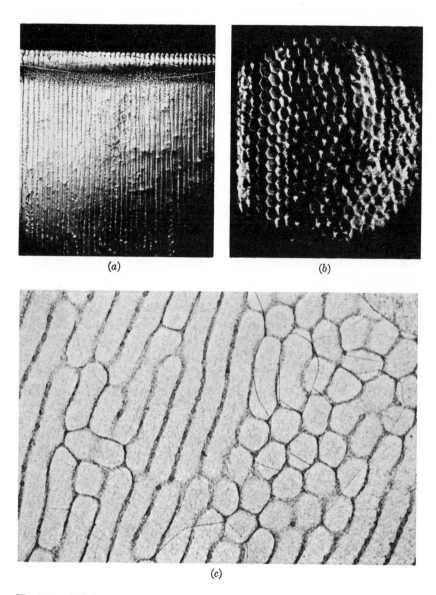

Fig. 5.29.   Cellular structure.   (a) View of top (free) surface of tin crystal (× 75);
(b) view of decanted interface of tin crystal (× 75); (c) less regular forms of cell.
Photographs (a) and (b) by R. W. Cahn, (c) by H. Biloni.

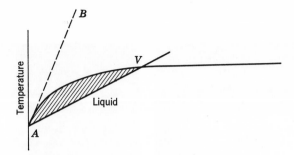

Fig. 5.30. Supercooling ahead of planar interface.

evolution of latent heat, it encountered a region where the supercooling
is sufficient only to provide the necessary kinetic driving force for
growth (Fig. 5.30). The protuberance grows until it reaches a steady
state growth condition; its convexity, as shown on page 138, leads
to a steady state growth temperature that is above that for the plane
interface, and a lower concentration of solute in the solid. The
rejection of solute produces lateral concentration gradients in addition
to the longitudinal ones that have been discussed (Fig. 5.31). The
accumulation of solute around the base of the protuberance retards
solidification in this region, and consequently the protuberance cannot
expand laterally. The convexity caused thereby in regions such as
P and Q triggers the development of similar protuberances around
the original one, and the result is an array of cells that have an approximately "close packed" structure, most of the cells having six
neighbors.

The qualitative predictions that can be made from this model
have been verified experimentally (11): (a) the cell surface is convex
towards the liquid; (b) the concentration of solute is higher at the

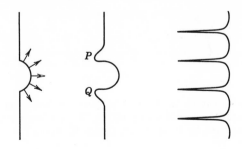

Fig. 5.31. Development of cellular interface.

cell walls than in the cell centers (15, 25), and is at a maximum at the cell corners (25); (c) cell formation can be suppressed by reducing the solute content, reducing the speed of growth, or increasing the temperature gradient to eliminate the region of supercooling (AV in Fig. 5.30).

Quantitative studies of cell formation. The cell formation theory of Rutter and Chalmers (11) predicts that cell formation can be suppressed by slow growth rate or a high temperature gradient; the theory of solute redistribution of Tiller, Jackson, Rutter, and Chalmers (3) leads to a quantitative prediction of the conditions that determine whether constitutional supercooling would exist ahead of a planar interface. If no constitutional supercooling could occur, then cells cannot be formed; if, on the other hand, constitutional supercooling is predicted, then its actual occurrence would be prevented by a departure from the conditions assumed in the calculation (i.e., a planar interface), and cells are formed.

If it is assumed that the solute has the steady state distribution, the critical temperature gradient that is required to prevent constitutional supercooling can be calculated as follows: if m is the slope of the liquidus line dT_E/dC_L, and T_0 is the equilibrium temperature for the pure metal, then

$$T_E = T_0 - mC_L$$

The equilibrium temperature at any point at a distance X in front of the interface is then given by

$$T_E = T_0 - mC_0 \left[1 + \frac{1 - k_0}{k_0} \exp\left(-\frac{R}{D} X \right) \right]$$

The actual temperature T at any point in the liquid can be expressed as

$$T = T_0 - m\frac{C_0}{k_0} + GX$$

(where G is the temperature gradient in the liquid in degrees per centimeter) since $T_0 - m(C_0/k_0)$ is the temperature of the interface (ignoring the supercooling required to provide the kinetic driving force).

Some values of these two functions, for the liquidus temperature and for the actual temperature, are plotted in Fig. 5.32. The length of the supercooled zone is given by the value of X at $T = T_E$, at which,

$$1 - \exp\left(\frac{R}{D} X \right) = \frac{G}{mC_0(1 - k_0)/k_0} X$$

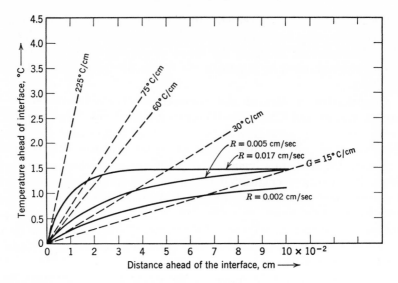

Fig. 5.32. Distribution of temperature and liquidus temperature ahead of an advancing planar interface, for selected values of R and G. (From Ref. 3, p. 434.)

The critical ratio of temperature gradient to growth rate is that at which the length of the supercooled zone is zero, or the slopes at the origin of the "liquidus" and "temperature" lines are equal. The condition is

$$\frac{G}{R} = \frac{mC_0}{D} \cdot \frac{1 - k_0}{k_0}$$

The qualitative experiments of Rutter and Chalmers (11) were followed by quantitative studies by Walton, Tiller, Rutter, and Winegard (26), by Tiller and Rutter (27), and by Plaskett and Winegard (28), who controlled both the speed of advance of the interface and the temperature gradient in the liquid by means of a system in which the temperatures of both source and sink were automatically controlled. A typical result of experiments of this kind is shown in Fig. 5.33, for lead with tin as a solute, which shows that the critical value of G/R is proportional to the concentration of solute, or G/RC_0 is constant. The quantitative value obtained for the critical value of G/R agrees very well with the value calculated from the known constants of the alloy system.

These experiments confirm that cell formation is a result of the instability produced by constitutional supercooling. It should be pointed out that the quantitative theory of constitutional supercooling is based on the assumption that the interface is planar and that there is no transverse motion of the solute; these conditions are satisfied so long as cells are not formed, but they break down as soon as cells are present.

Since constitutional supercooling cannot exist until an enriched (or depleted) boundary layer has been formed, it follows that cells should not form immediately when solidification begins, even if uniform values of G and R are immediately established. The critical condition will not be met until the initial transient has reached the stage at which the concentration gradient at the interface has a value such that the gradient of the liquidus temperature exceeds that of the actual temperature. It follows that there must be an "incubation distance" for the formation of cells; owing to the difficulty of immediately establishing constant values of R and G, there are no measurements of the incubation distance; however, it has been shown

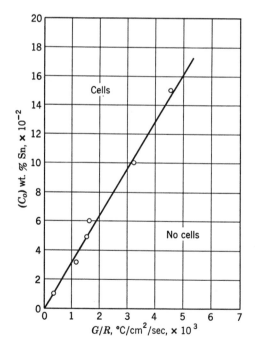

Fig. 5.33. Conditions for cellular solidification. (From Ref. 29, p. 250.)

(3) that the incubation distance Z should be given by

$$Z = \frac{D}{k_0 R} \ln \left[1 - \frac{GD}{C_0 R m (1 - k_0)/k_0} \right]$$

This is of the order of one millimeter for a typical experimental case.

It was pointed out by Rutter (29) that the critical condition for the formation of cells could be used as a method of determining the purity of a metal. If there is only a single solute of known k_0, a direct determination of the values of G and R at which cells are just formed allows C_0 to be calculated from

$$C_0 = \frac{k_0}{1 - k_0} \cdot \frac{D}{m} \cdot \frac{G}{R}$$

Rutter shows that, for experimentally convenient values of G and R, C_0 can be determined when it is between about 6×10^{-4} and 2×10^{-1} weight per cent.

If there is more than one solute, it is still possible to obtain information about the purity; but the information is restricted to an assessment of the combined effects of the solutes; the measured value of G/R is then equal to $\Sigma\{C_0^1 \cdot [(1 - k_0^1)/k_0^1] \cdot (m^1/D)\}$ where C_0^1, k_0^1 and m^1 refer to the concentration and phase diagram characteristics of solute 1, etc. It is also necessary to assume that the solution is so dilute that there is no interaction between solutes.

Considerable attention has been devoted to the problem of the geometry of the cellular interface; this includes three distinct aspects: (a) size of cells, (b) direction of cellular growth, (c) shape of cell surface. For clarity it is desirable to separate the qualitative observations, about which there is little ambiguity, from the theoretical discussion, which is not entirely satisfactory.

Geometry of cells. The major qualitative observation on cell size is that the size increases with decreasing rate of growth, but is apparently independent of the temperature gradient; Bolling and Tiller (30), in a mathematical discussion of the morphology of the cellular interface, show that very rough agreement with experimental data of Boĉek, Kratochvil and Valoual (31) can be obtained if it is assumed that the cell size is determined by the distance through which the solute in the enriched layer can diffuse laterally before the interface reaches it. The numerical values that are predicted, however, are very sensitive to the detailed assumptions that are made, and it is not surprising that good agreement has not been obtained. The direction of growth is more complicated, and has been studied by Teghtsoonian and Chalmers (32) and by Atwater and Chalmers (33), who, in the course of investigations on lineage structures (see Chapter 2) studied effects now known to be directly related to the direction of growth of the cellular structure. These studies, mainly on tin and on lead, showed that the direction of growth of the cells, as revealed by their walls, is not necessarily perpendicular to the mean interface, but deviates toward the nearest "dendrite direction" if this does not coincide with the normal to the interface. The extent to which the growth direction deviates depends on the speed, impurity content, and the inclination of the dendrite direction to the growth direction; the deviation is least at low speeds, and is always less than half of the deviation of the dendrite direction. The deviation increases with increase of impurity content. Some results obtained by Atwater are shown in Fig. 5.34.

There is no entirely satisfactory explanation for the direction of cellular growth; it appears to be certain that the deviation of the growth direction from that of heat flow must be related to an asymmetry in the shape of cells themselves, and this must in turn be caused by anisotropy of the growth rate, a feature of the growth of metallic crystals from the melt that has already been invoked to explain the direction of dendrite growth (Chapter 4).

The experimental evidence on the shape of cells has been somewhat misleading. Elbaum and Chalmers (34) studied the topography of interfaces whose growth had been interrupted by "decanting," that is, rapid withdrawal from the melt, by means of a device which caused

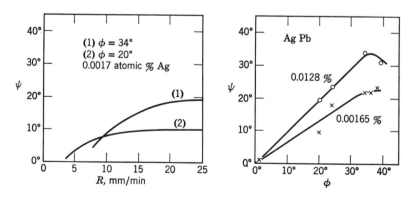

Fig. 5.34. Deviation of cellular direction ψ as a function of speed R orientation ϕ and solute. (From Ref. 33, p. 212.)

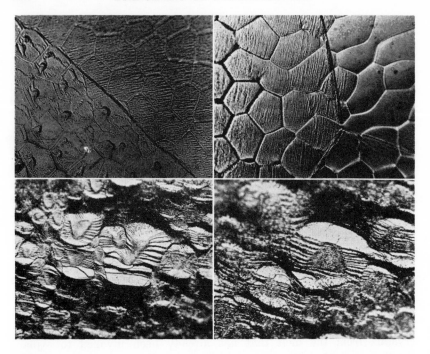

Fig. 5.35. Appearance of decanted cellular interface. (From Ref. 12, p. 522.)

the crystal to move away from the melt with an acceleration several
times that of gravity. Typical photomicrographs of cellular interfaces are shown in Fig. 5.35 in which it is seen that the surface is
characterized by a stepped or terraced structure. Elbaum showed that
the surface is terraced when it is within 10° of a {100} plane, or
within 20° of a {111} plane. Observations of this kind were accepted
as evidence that the interface was terraced during growth, and this
was used as the basis for speculation about the reason for the growth
direction of the cells. However, Chadwick (35) has shown that a
film of liquid is retained on a decanted surface and that the solidification of this layer, after withdrawal from the liquid, can produce the
terraced structure that is observed. The presumption that the terraces are formed after withdrawal is supported by the appearance of
"tear drops" on the solidified interface, as in the lower left-hand
corner of Fig. 5.35. It is concluded that there is no evidence that
terraces are present on the cell surfaces during growth, and it is
fruitless to examine the theoretical structure that was based on their
presumed existence.

The experimental observations are also somewhat misleading on the over-all shape of the cells; the photomicrographs of "decanted" interfaces suggest that the cell profile is as sketched in Fig. 5.35a, the cell walls apparently being essentially flat and of considerable thickness. Other evidence suggests, however, that the real shape of the cells is as shown in Fig. 5.36b, the walls extending relatively far back into the crystal. The evidence that supports this view is (a) that pores, probably due to shrinkage of the solidifying liquid, often occur in the cell walls (36). This would not occur if the shape were that shown in Fig. 5.36a, but the trapping of some liquid, and its subsequent shrinkage to form pores, would be expected if the shape is that of Fig. 5.36b; (b) it is often observed that a grain boundary follows cell walls during and immediately after growth (as revealed by the interface terrace structure) but that it moves to a position of smaller area while the terrace structure is forming. The explanation is that the boundary is constrained to remain in the cell walls while the interface has the form of Fig. 5.37a because any lateral motion would require the boundary to extend forward to the cell surface (Fig. 5.37b). When the liquid in the cell wall regions has solidified, the boundary can move laterally without first increasing its area to any substantial extent (Fig. 5.37b).

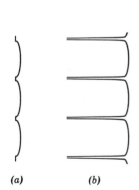

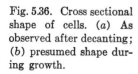

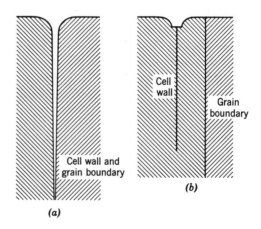

Fig. 5.36. Cross sectional shape of cells. (a) As observed after decanting; (b) presumed shape during growth.

Fig. 5.37. Relative positions of grain boundary and cell wall. (a) During growth; (b) after decanting.

5.7   Cellular Dendrites

One of the characteristics of the conditions in which cellular structures are produced is that there is a positive temperature gradient in the liquid, so that even in the absence of cells, the zone of constitutional supercooling would be very limited. If the temperature gradient is reduced, the zone of constitutional supercooling is extended, and the cells may change in character to such an extent that they acquire some of the characteristics of dendrites. When this occurs, the cell tips have a square pyramidal shape, instead of a rather flat curved "dome"; they form a square, instead of hexagonal, array, and they grow in the characteristic dendrite direction. They will be referred to hereafter as "cellular dendrites." The branches are nearly or quite continuous, forming "webs" in the {100} planes that are perpendicular to the main direction of growth. The structure is shown diagrammatically in Fig. 5.38a, and photographically in Fig. 5.38b. It was first described by Northcott (37) and has recently been studied by Flemings and Bowers (38) and by Biloni and Chalmers (39). This morphology is distinct from both the cellular structure and "free" dendritic growth, in which each dendrite grows independently of its neighbors. It is probable that the "cellular dendritic" type of growth occurs when the temperature gradient in the liquid is small, but positive, so that the latent heat of fusion is conducted into the solid, while the rejected solute diffuses outward. The distinction from the cellular structure arises because when the temperature gradient is sufficiently small, the crystallographic factors are able to control the shape and growth direction, as in the case of dendrites in a pure metal.

The alignment of the dendrites to form a square array in this kind of dendritic growth may arise from the fact that if the branches join to form webs, they provide a better conducting path for heat flow from the liquid to the crystal than would exist if the branches were separate. It is also possible, although it has not been demonstrated, that a web in (010) plane between two (100) dendrites is a stable growth form. It has been suggested by Flemings (40) that the formation of the "webs" may be a result of the fact that {100} surfaces should be slower growing than any other surfaces except {111}, and that, therefore, plates bounded by {100} surfaces are relatively stable. The distinction between "free dendrites" and "cellular dendrites" has been appreciated only very recently, and most of the work on the conditions for dendritic solidification in alloys relates to what are here described as cellular dendrites.

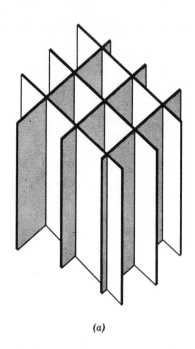

(a)

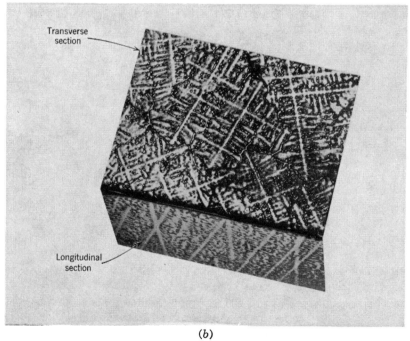

Transverse
section

Longitudinal
section

(b)

Fig. 5.38. Cellular-dendritic structure. (Photograph by T. Bowers.)

165

The criterion that determines whether growth is of the "cellular" or the "cellular dendritic" form has been studied experimentally by Morris, Tiller, and Rutter (41) who showed that composition, temperature gradient in the liquid, and rate of growth were all significant, and by Tiller and Rutter (27) who showed that the parameter $G/R^{1/2}$ appears to play the same part in relation to the transition from cellular to cellular dendritic solidification as G/R to the onset of cellular growth. It was, further, shown by Holmes, Rutter, and Winegard (42) and confirmed by Plaskett and Winegard (43) that the critical value of $G/R^{1/2}$ is proportional to C_0/k_0, or that the condition for cellular dendritic growth is $G/R^{1/2} \lesssim AC_0/k_0$, where A varies somewhat from one system to another, and depends, by a factor of two, upon whether the first sign of "breakdown" is looked for, or whether the fully developed cellular dendritic structure is regarded as the criterion.

The criterion for the transition from the cellular to the cellular dendritic type of growth appears to demand that there is a critical value of $G/R^{1/2}$ which is proportional to the concentration of solute at the interface (given by C_0/k_0). The significance of $G/R^{1/2}$ can be explained as follows: the cell size Z, which is determined by a diffusion distance, should be inversely proportional to the square root of the velocity R, or $Z = aR^{-1/2}$. The thickness of the supercooled layer t is given by

$$1 - \exp\left(-\frac{R}{D}t\right) = \frac{G}{mC_0(1 - k_0)k_0}$$

when substituting $a/R^{1/2}$ for t,

$$1 - \exp\left(-\frac{Ra}{DR^{1/2}}\right) = \frac{aG}{R^{1/2}} \cdot \frac{1}{mC_0(1 - k_0)/k_0}$$

or

$$1 - \exp\left(-\frac{R^{1/2}a}{D}\right) = \frac{aG}{R^{1/2}} \cdot \frac{1}{mC_0(1 - k_0)/k_0}$$

if the value of $\exp(-R^{1/2}a/D)$ is small compared with 1, then the criterion $Z = t$ is satisfied for a value of $G/R^{1/2}$ that is proportional to C_0 and to $1/k_0$.

Cellular-dendritic growth differs from cellular growth in that, because the depth of the supercooled zone is greater, the cells become more pointed, with the result that the factors that determine the direction of growth of dendrites are able to exercise control; that is, growth is relatively slow in four {111} directions, accounting for both the pyramidal morphology and the direction of growth. Cellular-dendritic growth occurs when the temperature of the liquid is determined largely or completely by that of the interface through which heat is extracted, and the cellular-dendritic growth front is the first stage of a single solidification process that terminates with the solidification of the last remaining liquid in the spaces between the "walls" of the structure. For the same reasons as in the cellular structure, the liquid between the first solidified material has a higher solute content and therefore a lower liquidus temperature; hence the interdendritic spaces remain liquid until the "tips" have advanced a substantial distance.

The rate of growth of the dendrite tips of the cellular-dendritic structure is controlled by the rate of heat extraction through the solid and therefore, by the rate of advance of the appropriate isotherm into the liquid. The shape of the freezing front, similarly, is controlled by the shape of the isothermal surface, which in turn depends on the geometry and thermal characteristics of the system.

It has recently been shown by Biloni and Chalmers (2) that a structure described as "pre-dendritic" characterizes the first part of a crystal to form in an alloy. There is a smooth transition between the pre-dendritic structure and the cellular dendritic structure which develops from it, as shown in Fig. 5.39.

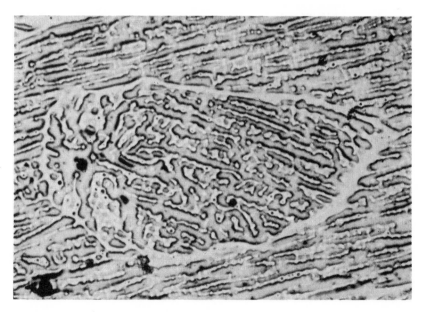

Fig. 5.39. Crystal showing free dendrites and cellular dendritic substructures.

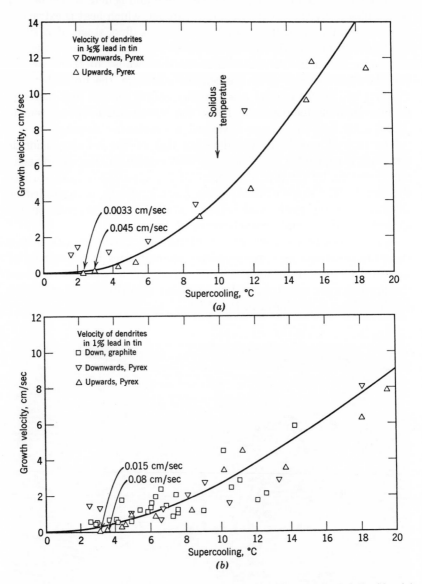

Fig. 5.40. Rate of growth of free dendrites of Pb-Sn in supercooled liquid. (a) Tin with ½ per cent lead, (b) tin with 1 per cent lead. (Data from Ref. 44.)

5.8   Free Dendritic Growth in Alloys

Cellular dendritic growth in alloys is, in a sense, intermediate between cellular growth, which is the stable alternative to a thin zone of constitutional supercooling, and free dendritic growth, which takes place by conduction of latent heat outward from the growing crystal into a supercooled melt. In free dendritic growth, the rate of advance of the growing points is determined only by the temperature and composition, and the shape of a crystal growing in this way is similarly determined by the local conditions of growth and not by any externally imposed temperature gradient. It appears that this type of growth is conducive to the formation of "branched rod" type dendrites, with a morphology similar to that of pure metals, while an imposed temperature gradient tends to produce the plate type of structure described as cellular dendritic. The reason for this difference may be that, while the branches of the free type of dendrite do not grow after they have "used up" the local supercooling, those of the cellular type do so, because of the continuous extraction of heat; the consequent thickening of the branches may lead to their coalescence as plates.

Free dendritic growth in alloys takes place in a manner very similar to that already described for pure metals. However, the rejection of solute presents an additional complication, which results in slower growth than would be found for the pure metal. The liquid in contact with the interface at the tip of a growing dendrite must have a higher solute content than that of the ambient liquid, and therefore its equilibrium temperature is reduced. This decreases the effective supercooling of the melt. The heat flow considerations that are valid for the pure metal should still apply if the corrected value of the equilibrium temperature of the interface is used. However, the solute diffusion equation that would give the equilibrium temperature itself depends on the rate of growth and the radius of the tip; the problem of predicting the growth rate in a supercooled alloy has not yet been solved; however, there have been two relevant experimental investigations, those of Orrok (44) on tin alloys and of Walker (45) on copper-nickel alloys. The results of these studies are shown in Figs. 5.40 and 5.41.

An interesting case of dendritic growth, in which the controlling process is probably diffusion, has been observed by Willens (46) in an electron microscope study of the crystallization of an alloy of 15% germanium, 85% tellurium, which had been cooled so rapidly (47) that an amorphous solid was produced. The heating due to the electron beam in the microscope was sufficient to allow crystallization to proceed; it is probable that the phase which crystallizes is different in composition from the matrix in which it is growing.

Spacing of dendrite arms. The spacing of dendrite arms in alloys that have solidified dendritically has been studied by Alexander and Rhines (48), Michael and Bever (49), and by Howarth and Mondolfo (50). As the results of these investigations are in general agreement, it is sufficient to summarize the most recent, in which a series of aluminum copper alloys were cooled unidirectionally at various rates, the temperature gradient as well as the rate of cooling being measured. Somewhat surprisingly, it was found that for a given alloy, the spacing of the dendrite arms correlated reasonably well with the rate of cooling (expressed in deg/sec) rather than with the rate of motion of the isotherms, expressed in centimeters per second. The spacings can be represented by

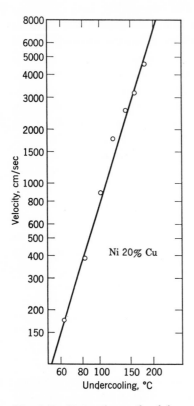

Fig. 5.41. Rate of growth of free dendrites in a supercooled melt of Ni-0.20% Cu. (From Ref. 45.)

$$S = A \exp (B \ln R + CM)$$

where R is the rate of cooling, M the molar fraction of copper, and A, B, and C the constants—of which, however, A and C have values that are different in hypo- and hyper-eutectic alloys. The dendrite arm spacing decreases with increasing cooling rate, explained on the basis that there is less time available for diffusion of solute, and also as the eutectic composition is approached, presumably because of the increasing amount of solute that is rejected as the composition moves further away from the pure metal. Howarth and Mondolfo concluded from these results that the spacing of the arms is controlled by diffusion and not by heat transfer, and that the spacing is determined by the characteristic thickness of the diffusion zone around a growing dendrite.

5.9   Nucleation of Crystals ahead of the Existing Interface

An additional consequence that can arise from constitutional super-cooling is the nucleation of new crystals in liquid which is at a higher temperature than the interface at which growth is taking place. The presence of nucleant particles of sufficient potency in the zone of con-stitutional supercooling is sufficient, and it is believed that one of the main difficulties encountered in growing single crystals from impure or alloy melts is that this combination of an enriched boundary layer and possible nucleants is likely to occur. The fact that increasing the temperature gradient and decreasing the speed improve the probability of success indicates that this source of "stray" crystals is a real one.

Nucleation in a constitutionally supercooled zone may also be im-portant in the solidification of a liquid alloy. Although it now ap-pears that supercooling is less important in the solidification of alloy castings and ingots than had been supposed (see Chapter 8), it can be responsible for the nucleation of crystals ahead of a dendritic or cellular dendritic interface. Plaskett and Winegard (43) have shown that when an alloy of aluminum and magnesium is solidified at a rate R in a temperature gradient G, nucleation of new crystals occurs ahead of an advancing interface when $G/R^{1/2}$ is less than a value that is roughly proportional to the magnesium content of the alloy. They explain this result on the basis that the value of $G/R^{1/2}$ determines the maximum supercooling ahead of the interface, and that nucleation occurs when the supercooling reaches the value at which nucleation occurs on the most potent nucleant particles that are present.

5.10   Types of Segregation

We will now summarize the discussion of the redistribution of solute in a single-phase alloy by describing the various types of segregation that can occur. All types of segregation are the result of the re-jection of solute (or of solvent) at the interface during solidification; the different types of segregation differ only in the direction, distance, and extent of the motion of the solute.

Normal segregation. This type of segregation is defined in terms of motion of solute parallel to the direction of solidification; if trans-verse concentration gradients occur, they are ignored and the variation of the average concentration, ideally in successive positions of the

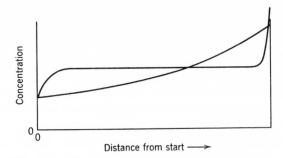

Fig. 5.42. Extreme cases of normal segregation.

interface, is taken to indicate the extent of normal segregation. It follows that the normal segregation of a solute can be represented by a single curve, relating the concentration of a solute to the distance from the start of solidification. The form of this curve depends upon the equilibrium distribution coefficient k_0, the rate R of solidification (or the time-distance relationship if R is not constant), and the amount of mixing by fluid motion. The extreme possible forms of the normal segregation curve are shown in Fig. 5.42. The actual segregation resulting from a solidification process depends very much on the geometry of the sample, as this influences both the rate of solidification and the amount of convection; it is not, therefore, possible to state general rules for predicting the result. However, Winegard (51) shows that, by combining the theory of Wagner (8) and some experimental work of Weinberg (52), it can be predicted that when the rate of solidification is below about 2 cm/hr, the segregation pattern will be dominated by mixing, while at speeds above 20 cm/hr the result will be essentially that to be expected from diffusion control; this is, as Winegard points out, an oversimplification, because the form of the interface, which depends on the extent of constitutional supercooling, has an important influence on the mixing process.

Grain boundary segregation. Segregation may occur at grain boundaries during the process of solidification; this is not to be confused with *equilibrium* segregation at grain boundaries, which is an adsorption type of phenomenon and is usually small in extent. When two crystals grow side by side (Fig. 5.43), the boundary between them forms a groove which, according to Tiller (53) extends a distance of the order of 10^{-3} cm behind the main interface. Tiller has examined the conditions under which marked transverse diffusion of solute, resulting in segregation at the boundary, should occur. The case considered is one in which cellular growth does not occur, and it is shown that there is no significant segregation to the boundary unless there is some constitutional supercooling, in which case cellular segregation would also occur. It has been shown by Biloni and Bolling (54) that there is marked segregation to the low-angle boundaries that are the characteristic feature of macromosaic or lineage structures.

Cellular segregation. It was clearly shown by Rutter and Chalmers (11) that after cellular solidification there is a higher concentration of solute at the cell walls than in the cells. This qualitative observation has been confirmed by numerous workers; the most thorough study is that of Biloni and Bolling (54), who, using refined metallographic techniques, showed that the degree of segregation to cell boundaries can be so large that, for example, the lead rich phase can be seen at cell boundaries in a tin–0.16 per cent lead alloy. This indicates that at the "terminal transient" region, where three neighboring cells grow toward each other, a fifteenfold increase in concentration may occur.

There is no satisfactory theory for predicting either the total amount of segregation at the cell boundaries or the maximum concentration that is reached during solidification; a recent paper by Kramer, Bolling, and Tiller (55) emphasizes the formidable nature of the problems that must be overcome before such a theory can be formulated. It has been shown that the segregation decreases, as a result of diffusion, during cooling down after solidification, and during any subsequent annealing process, but it is possible, as suggested by Atwater and Chalmers (33) that dislocations associated with the solute at the cell walls may tend to stabilize it and limit the homogenizing effect of annealing.

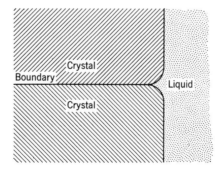

Fig. 5.43. Conditions for grain boundary segregation.

Fig. 5.44. Distribution of copper in an Al-4.5% Cu alloy. (From Ref. 56.)

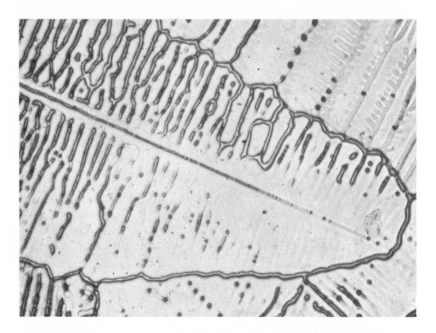

Fig. 5.45. Segregation in Al-Cu alloy. (Photograph by H. Biloni.)

Dendritic segregation. Almost all the work on microsegregation in alloys has been on cellular-dendritic structures, in which, it has been shown by Flemings (56), there is very strong segregation; Fig. 5.44, for example, is a microradiograph showing the distribution of copper in an aluminum 4.5 per cent copper alloy. It is interesting that the copper, which appears light in the illustration, is most concentrated close to the plates, which are relatively free from copper. Biloni and Chalmers (39) have used Biloni's epitaxial film technique to show the segregation in aluminum-copper alloys; Fig. 5.45 is an example. The observations that have been made on segregation in cellular dendrites can all be accounted for with the following assumptions (56). (1) Liquid is effectively "trapped" early in the development of the structure; there is therefore no long-range segregation except by fluid motion in the region at and ahead of the "tips" of the growing crystals. (2) Solute is rejected according to the laws discussed above. (3) Diffusion of solute within each local region of the cellular dendritic structure is sufficient for the liquid in it to be effectively uniform at any time. Quantitative work on the actual distribution of solute has

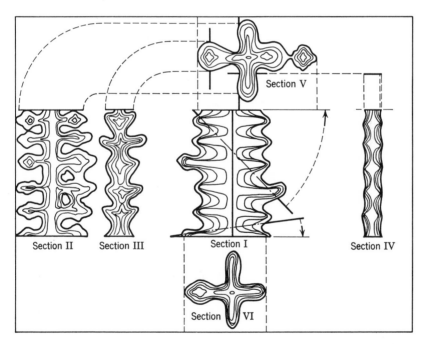

Fig. 5.46. Quantitative results on distribution of solute in columnar dendritic crystal. (From Ref. 57.)

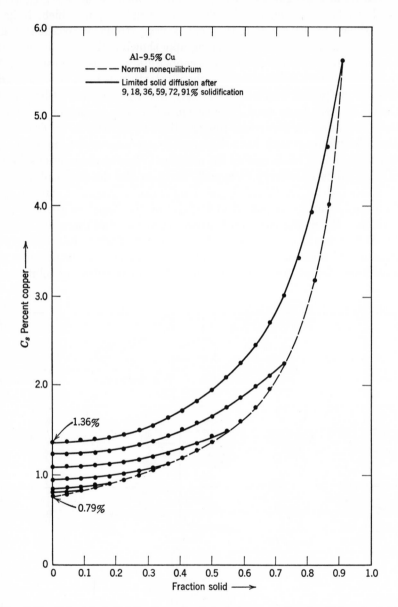

Fig. 5.47. Calculated distribution of solute between walls of cellular dendritic structure. (From Ref. 56.)

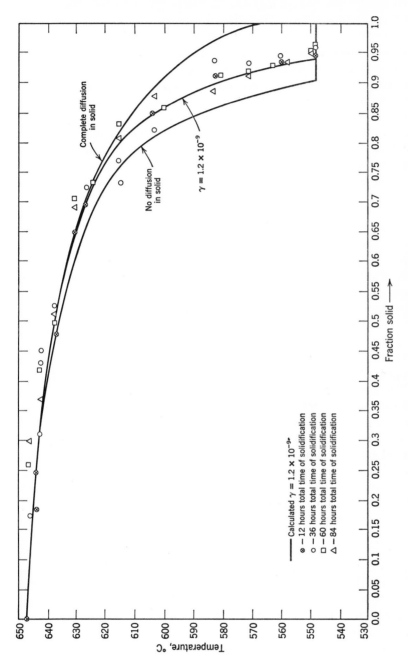

Fig. 5.48.   Comparison between calculated and experimental distributions of solute.   (From Ref. 56.)

been carried out by Flemings and Kattamis (57) who used the electron microprobe technique, with which they were able to establish solute isoconcentration lines such as those shown in Fig. 5.46, from which it is apparent that the plate-like structure is in this case built up of a main stem and coalesced branches. The extremely complicated geometry of the cellular structure makes it very difficult to form realistic theoretical predictions with which to compare the experimental results. Flemings and Brody (56), however, have shown that good agreement with the electron-probe results can be obtained from calculations based on a very simple model, in which solidification is regarded as taking place inward from two plane "walls" whose spacing is equal to that of the cell walls. The calculated distribution using

$$C_s = k_E C_0 (1 - g)^{k_0 - 1}$$

is shown in Fig. 5.47 for some aluminum copper alloys; it is of interest that the speed of solidification does not appear explicitly in this solution, because the cell wall spacing depends upon speed. The distribution of solute, therefore, is independent of speed if it is plotted in the dimensionless form of Fig. 5.47. The maximum concentration is defined by the eutectic composition and is therefore independent of the rate of solidification. Flemings and Brody (56) have compared experimental results with theoretical values based on various assumptions of the amount of diffusion that occurs in the solid during solidification. The results of such a comparison are shown in Fig. 5.48.

The small scale of the cellular dendritic structure, and the large differences in concentration between the initial and terminal transient regions, give rise to very steep concentration gradients in the alloy as it solidifies; it is therefore to be expected that some homogenization should take place by diffusion while the alloy is cooling down after solidifying. Flemings and Poirier (58) have shown that the amount of homogenization during cooling can be predicted with reasonable accuracy if the initial distribution is calculated in the manner discussed above. A comparison between theory and experiment is shown in Fig. 5.49.

Inverse segregation. Inverse segregation occurs when a solute that is rejected during solidification is present at a higher concentration in regions that solidified earlier than in those that solidified later; in other words, solute moves in a direction opposite to that of normal segregation. The literature contains numerous discussions of the causes of inverse segregation (see, for example, Hanson and Pall-Walpole (59)), but recent studies indicate a consensus of opinion, supported by experiment (60, 61, 62, 63), that inverse segregation

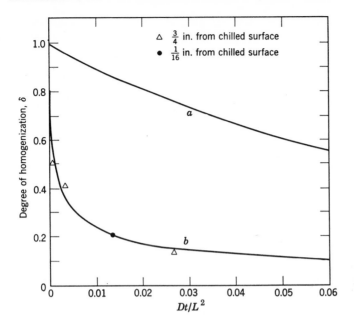

Fig. 5.49.   Experimental and calculated results for homogenization during cooling. (From Ref. 58.)

takes place in the following way (originally proposed by Scheil (64)). The shrinkage that in most alloys accompanies solidification may cause motion of the most "solute enriched" liquid in a direction opposite to that of the general solidification front.   Figure 5.50 illustrates this.   The material to the left of the line OP is assumed to be com-

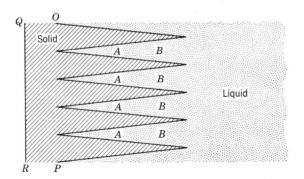

Fig. 5.50.   Mechanism of inverse segregation.

pletely solidified; as the regions AA solidify, shrinkage takes place, causing the solute enriched liquid at BB to move toward the left. Thus there is net transfer of solute from right to left. The effect is larger than might be expected because of the highly enriched composition of the liquid in the terminal transient condition, i.e., when there is only a small amount remaining in a given region. The general validity of this theory is supported by the observation of Adams (65) that inverse segregation does not occur in alloys that expand on solidification.

Kirkaldy and Youdelis (63) have developed the quantitative approach originally proposed by Scheil to the theory of inverse segregation outlined above. Their analysis is based on the following assumptions:

1. There are no shrinkage voids; that is to say, the liquid is able to flow outward to compensate completely for the shrinkage.

2. The liquid in the interdendritic region is always of uniform (though changing) composition, and it is always in equilibrium with the solid that has just been deposited on the dendrites. The dendrites, however, are not of uniform composition, since they are formed from a liquid of continuously changing composition. The theory describes an alloy system in which enrichment of the liquid is limited by a eutectic composition; the case considered by Kirkaldy and Youdelis was that of the aluminum copper system, of which the relevant part of the phase diagram is shown in Fig. 5.51.

The agreement between theory and experiment is shown in Fig. 5.52.

The case of the aluminum zinc system (Fig. 5.53), studied by Youdelis and Colton (66), is somewhat more complicated because the assumption of constant k_0 made by Kirkaldy and Youdelis can no longer be regarded as correct, and stepwise integration is used. The result of the computation is a graph of segregation as a function of zinc concentration which takes the form of the continuous line in Fig. 5.54. The agreement between theory and experiment is shown in the same Figure.

A much more drastic form of inverse segregation occurs when liquid is sucked out through the boundary QR (Fig. 5.50) (the first region to solidify) by a difference of pressure arising from the separation of the metal from the mold. The liquid which has exuded solidifies on the surface of the metal between it and the mold, in the form of small "drops," known as "sweat," which are usually much harder than the neighboring metal because of their high alloy content. The theory discussed above assumes the absence of surface exudation.

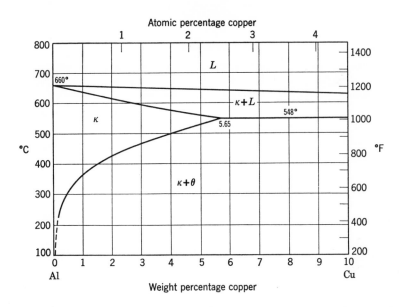

Fig. 5.51. Part of the phase diagram for the Cu-Al system. (From *Metals Handbook*, American Society for Metals, Cleveland, 1948, p. 1160.)

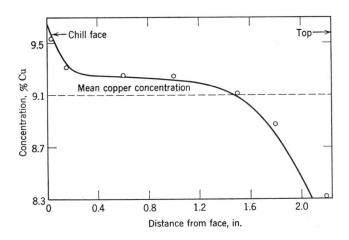

Fig. 5.52. Comparison between experimental and calculated inverse segregation in Al-Cu. (From Ref. 63.)

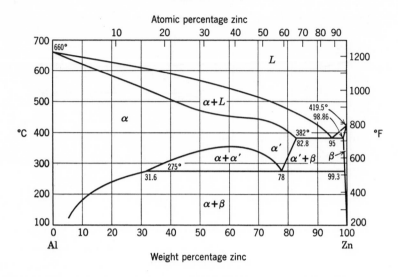

Fig. 5.53. Part of phase diagram for the Al-Zn system. (From *Metals Hand-book*, American Society for Metals, Cleveland, 1948, p. 1167.)

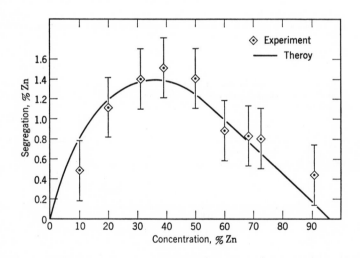

Fig. 5.54. Predicted and measured inverse segregation in Al-Zn. (From Ref. 66.)

Coring and intercrystalline segregation. The discussion of normal segregation (page 171) was developed from a macroscopic point of view; the distances over which normal segregation are important are usually large compared with the individual crystals of which the alloy is composed. Each crystal, however, is likely to exhibit normal segregation on a microscopic scale; the earliest part of it to form has a lower concentration of solute (unless $k_0 > 1$) than the later parts; this is called "coring"; the highest solute content, locally, is usually to be found in the terminal regions where two or more crystals grow toward each other and finally form a grain boundary.

5.10   Gravity Segregation

When the mass of liquid is sufficiently large, convection causes motion of the solute away from the limit of the boundary layer. If the solute changes the density of the liquid, it may set up a convection current that carries it toward the top or the bottom of the space in which the liquid can move. This is a type of transverse segregation that is much more significant in large, real systems than in the idealized ones discussed in this chapter.

References

1. K. A. Jackson, *Can. J. Phys.*, **36**, 683 (1958).
2. H. Biloni and B. Chalmers, Unpublished.
3. W. A. Tiller, K. A. Jackson, J. W. Rutter, and B. Chalmers, *Acta Met.*, **1**, 428 (1953).
4. J. Frenkel, *Kinetic Theory of Liquids*, Oxford University Press, London, 1946.
5. W. Seith, *Diffusion in Metallen*, Springer-Verlag, Berlin, 1946, p. 285.
6. A. Kohn and J. Philibert, *Rev. Metallugie*, April, 1960.
7. V. G. Smith, W. A. Tiller, and J. W. Rutter, *Can. J. Phys.*, **33**, 723 (1955).
8. C. Wagner, *Trans. AIME*, **200**, 154 (1954).
9. W. G. Pfann, *Zone Melting*, John Wiley & Sons, New York, 1958.
10. J. A. Burton, R. C. Primm, and W. P. Slichter, *J. Chem. Phys.*, **21**, 1987 (1953).
11. J. W. Rutter and B. Chalmers, *Can. J. Phys.*, **31**, 15 (1953).
12. B. Chalmers, *Trans. AIME*, **200**, 519 (1956).
13. A. I. Landau, *POMM*, **6**, 132 (1958).
14. J. A. Burton, E. D. Kolb, W. P. Slichter, and J. D. Struthers, *J. Chem. Phys.*, **21**, 1991 (1953).
15. M. T. Stewart, R. Thomas, K. Wauchope, W. C. Winegard, and B. Chalmers, *Phys. Rev.*, **83**, 657 (1951).
16. N. Albon, *J.A.P.*, **33**, 2912 (1962).
17. H. C. Gatos, A. J. Strauss, M. C. Lavine, and T. C. Harmon, *J.A.P.*, **32**, 2057 (1961).

18. M. Smialowski, *Z. Metallk.*, **29**, 133 (1937).
19. L. Graf, *Z. Phys.*, **121**, 73 (1943).
20. B. Chalmers and R. W. Cahn, Unpublished work, 1947.
21. A. J. Goss and S. Weintraub, *Nature*, **167**, 349 (1951).
22. R. B. Pond and S. W. Kessler, *J. Met.*, **3**, 1156 (1951).
23. A. Prince, *J. Met.*, **4**, 1187 (1952).
24. F. Rosi, *J. Met.*, **5**, 1661 (1953).
25. H. Biloni, *Can. J. Phys.*, **39**, 1501 (1961).
26. D. Walton, W. A. Tiller, J. W. Rutter, and W. C. Winegard, *Trans. AIME*, **203**, 1023 (1955).
27. W. A. Tiller and J. W. Rutter, *Can. J. Phys.*, **34**, 96 (1956).
28. T. S. Plaskett and W. C. Winegard, *Can. J. Phys.*, **37**, 1555 (1959).
29. J. W. Rutter, *Liquid Metals and Solidification*, American Society for Metals, Cleveland, 1958, p. 250.
30. G. F. Bolling and W. A. Tiller, *J.A.P.*, **31**, 2040 (1960).
31. M. Boček, P. Kratochvil, and M. Valouchi, *Czech. J. Phys.*, **8**, 557 (1958).
32. E. Teghtsoonian and B. Chalmers, *Can. J. Phys.*, **39**, 370 (1951); **30**, 188 (1952).
33. H. Atwater and B. Chalmers, *Can. J. Phys.*, **35**, 208 (1957).
34. C. Elbaum and B. Chalmers, *Can. J. Phys.*, **33**, 196 (1955).
35. G. Chadwick, *Acta Met.*, **10**, 1 (1962).
36. P. E. Doherty and R. S. Davis, *Trans. Met. Soc. AIME*, **221**, 737 (1961).
37. L. Northcott, *J. Inst. Met.*, **65**, 205 (1939).
38. M. C. Flemings and T. Bowers, *Private Communication*.
39. H. Biloni and B. Chalmers, Unpublished work.
40. M. C. Flemings, *Private Communication*.
41. J. Morris, W. A. Tiller, J. W. Rutter, and W. C. Winegard, *Trans. ASM*, **47**, 463 (1955).
42. E. L. Holmes, J. W. Rutter, and W. C. Winegard, *Can. J. Phys.*, **35**, 1223 (1957).
43. T. S. Plaskett and W. C. Winegard, *Trans. AIME*, **51**, 222 (1950).
44. T. Orrok, Thesis, Harvard, 1958.
45. J. L. Walker, *Private Communication*.
46. R. H. Willens, *J.A.P.*, **33**, 3269 (1962).
47. P. Duwez, R. H. Willens, and W. Klenert, *J.A.P.*, **31**, 1136 (1960); and *Nature*, **187**, 869 (1960).
48. B. H. Alexander and F. N. Rhines, *Trans. AIME*, **188**, 1267 (1950).
49. A. B. Michael and M. B. Bever, *Trans. AIME*, **188**, 47 (1950).
50. J. A. Howarth and L. F. Mandolfo, *Acta Met.*, **10**, 1037 (1962).
51. W. C. Winegard, *Metallurgical Rev.*, **6**, 21 (1961).
52. F. Weinberg, *Trans. Met. Soc. AIME*, **221**, 844 (1961).
53. W. A. Tiller, *J.A.P.*, **33**, 3206 (1962).
54. H. Biloni and G. F. Bolling, *Trans. Met. Soc. AIME*, **227**, 1351 (1963).
55. J. J. Kramer, G. F. Bolling, and W. A. Tiller, *Trans. Met. Soc. AIME*, **227**, 374 (1963).
56. M. C. Flemings and H. Brody, *Private Communication*.
57. M. C. Flemings and L. Kattamis, *Private Communication*.
58. M. C. Flemings and D. Poirier, *Private Communication*.
59. D. Hanson and W. T. Pell-Walpole, *Chill Cast Tin Bronzes*, Edward Arnold, London, 1951.

60. S. Muromachi, *Nippon Kinzoku Gakkaishi*, **17**, 397 (1953).
61. W. C. Winegard, *Trans. Am. Foundryman's Soc.*, **61**, 352 (1953).
62. A. C. Simon and E. L. Gauss, *J. Electrochem. Soc.*, **101**, 536 (1954).
63. J. S. Kirkaldy and W. V. Youdelis, *Trans. Met. Soc. AIME*, **212**, 833 (1958).
64. E. Scheil, *Metallforschung*, **2**, 69 (1947).
65. D. E. Adams, *J. Inst. Met.*, **75**, 805 (1948–1949).
66. W. V. Youdelis and D. R. Colton, *Trans. Met. Soc. AIME*, **218**, 628 (1960).

6

Polyphase
Solidification

The discussion in Chapter 5 of the redistribution of solute during solidification was based on the implicit assumption that the whole of the solute could be accommodated in solution in a single solid phase; that is, that the relevant solid-liquid equilibrium relationship could be represented by a single liquidus line and the corresponding solidus line. There are, however, many important cases in which more than one solid phase is formed; in binary alloy systems the number of phases formed is limited to three, of which one may be gaseous, and in more complex systems the number may be larger. The principles that govern polyphase solidification have been investigated almost exclusively with binary systems, and the discussion will therefore be mainly related to such cases.

6.1 Evolution of a Gas During Solidification

Gas metal equilibria. In considering the behavior of metallic systems in which the relevant phases are liquid and solid, it is nearly always permissible to ignore the pressure to which the system is subjected; in fact the only case in which pressure appears to be relevant is for the consideration of nucleation under transient pressures resulting from cavitation (see Chapter 3). However, when one of the phases is a gas (other than the vapor of the metal) the pressure has an important influence on the equilibrium relationships. The phase diagram for a system containing a gas phase as well as liquid and solid phases requires pressure as a variable, in addition to temperature and composition. It is seldom necessary, however, to consider the phase diagram as a whole, because the variation in melting point, or in the liquidus and solidus temperatures, due to the presence of the gas is always small, unless a new phase, such as a metal-gas compound, is formed. The important characteristic of gas-metal systems

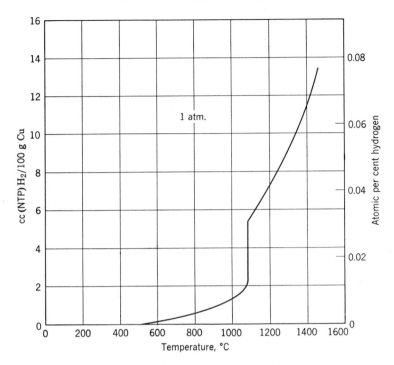

Fig. 6.1. Solubility of hydrogen in copper. (From M. Hansen, *Constitution of Binary Alloys,* McGraw-Hill, New York, 2nd ed., 1958, p. 587.)

from the present point of view is the solubility of the gas in the metal; that is, the concentration of gas in solution in the solid or the liquid that is in equilibrium with the same gas at a known pressure or partial pressure. For many purposes, a diagram relating solubility with temperature, for a single pressure (for example, a pressure of one atmosphere) will convey the significant information. Some typical solubility-temperature diagrams are given in Figs. 6.1, 6.2, and 6.3. It will be apparent from these diagrams that the solubility of a gas is, in general, less in the solid than in the liquid; this is equivalent to the statement that the value of k_0 is less than unity.

A case which is very important technologically is the solubility of hydrogen in iron; it is stated by Smialowski (1), that at atmospheric pressure, 100 grams of liquid iron dissolves 27 cubic centimeters of hydrogen, while the corresponding figure for the solid at the melting point is 13 cubic centimeters.

Bubble formation. When a metal or alloy containing a gas (such as oxygen, nitrogen or hydrogen) in solution solidifies, gas will

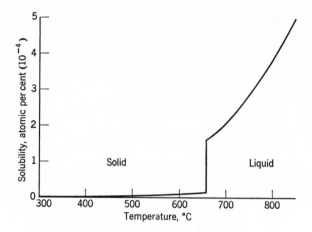

Fig. 6.2. Solubility of hydrogen in aluminum.

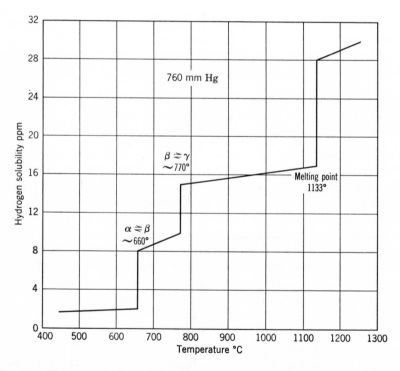

Fig. 6.3. Hydrogen-uranium phase diagram. (From M. Hansen, *Constitution of Binary Alloys*, McGraw-Hill, New York, 2nd ed., 1958, p. 803.)

be rejected at the interface, exactly in the same way as any other solute for which the $k_0 < 1$. If the gas in the liquid was already saturated (i.e., if it was at the limit of solubility for the prevailing pressure, at the melting point) then it becomes supersaturated as soon as an enriched layer begins to form at the interface. This means that there is more gas in solution than the equilibrium amount, and therefore there is a thermodynamic driving force tending to reduce the gas content. The amount of gas in solution may decrease either by escape of the gas at a free surface, if one is accessible within the range in which the gas can diffuse in the liquid, or by the formation of a gas bubble in the liquid. The formation of a gas bubble requires nucleation, which may be either homogeneous or heterogeneous. The condition for nucleation of a gas bubble is similar to that for nucleation of a solid phase, except that the effect of the pressure of the gas on its free energy must be taken into account. It follows that the nucleation of a bubble of a dissolved gas in a liquid is, at least formally, similar to the other nucleation problems discussed in Chapter 3; and for the reasons given there, heterogeneous nucleation can occur under conditions that would not permit homogeneous nucleation.

It can be shown, as follows, that a solid-liquid interface should not be an effective nucleant for a bubble; let the specific free energies of the solid-liquid, solid-gas and liquid-gas interfaces be σ_{SL}, σ_{SG}, and σ_{LG}, respectively, and consider the equilibrium of a bubble that has formed at an interface, as shown in Fig. 6.4. Equilibrium corresponds to a value of θ less than $180°$, when

$$\sigma_{SL} = \sigma_{SG} + \sigma_{LG} \cos \theta$$

$$\cos \theta = \frac{\sigma_{SL} - \sigma_{SG}}{\sigma_{LG}}$$

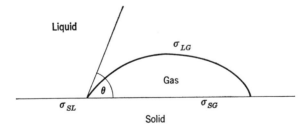

Fig. 6.4. Solid-liquid interface as nucleant for a gas bubble.

It is known that σ_{SG} is much greater than σ_{SL}, and that σ_{SG} is likely to be somewhat larger than σ_{LG}; the value of cos θ is therefore negative and may be less than -1, in which case the surface energy of the bubble is increased by contact with the solid-liquid interface.

It is observed, however, that gas bubbles are formed at solid-liquid interfaces. This location is in part due to the fact that the gas concentration would be highest there during solidification; but it may also be due to the fact that any re-entrant in the interface, such as a cell wall, grain boundary, or interdendritic space, would have an even higher gas content because of lateral segregation, as shown in Fig. 6.5. In regions such as A, the terminal transient condition (see Chapter 5) is entered as the 2 walls approach each other; the concentration of solute (in this case, gas) rises far above the value C_0/k that would occur at a flat interface, and which might not be sufficient to cause nucleation of gas bubbles.

Experimental observations leave no doubt that gas bubbles are in fact nucleated during solidification, when transport of gas away from the interface by diffusion is not sufficiently fast, in terms of the rate of rejection at the interface, to hold the gas content below the nucleation level. The subsequent behavior of the bubbles depends upon whether they float away from the surface or remain attached. If a bubble escapes from the region where it was nucleated, it may float to the surface of the melt, or it may be trapped by other crystals, in which case its subsequent behavior is similar to that of a bubble that remained at its point of origin. If the bubble is trapped where it

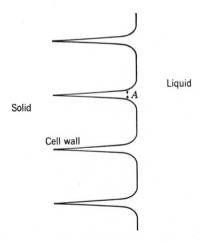

Fig. 6.5. Conditions for the nucleation of gas bubbles in cell walls.

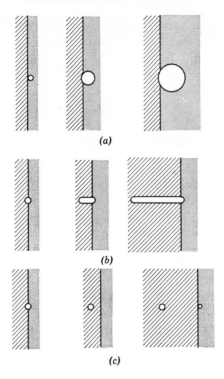

Fig. 6.6. Effect of speed of growth of a bubble on its shape and size. (a) Slow growth, (b) intermediate speed, (c) fast growth.

forms, it immediately becomes a "sink" into which gas from the neighboring supersaturated liquid can escape. This sets up a concentration gradient which causes gas to diffuse to the bubble from the surrounding liquid. The bubble therefore grows; but while it is doing so, the interface continues to advance. The relative rates of growth of the bubble and of advance of the interface determine whether the bubble will increase in diameter, remain of constant diameter but increase in length, or be overgrown by the solid. These three possibilities are shown schematically in Figs. 6.6a, 6.6b, and 6.6c.

A bubble that starts by growing (Fig. 6.6a) may reach a size at which the conditions of Fig. 6.6b are satisfied, and a cylindrical bubble becomes stable. The stability of the cylindrical bubble at intermediate speeds, and its breakdown at high and at low speeds, has been demonstrated for water containing dissolved air by Chalmers and Newkirk (2). The cylindrical type of bubble, sometimes referred

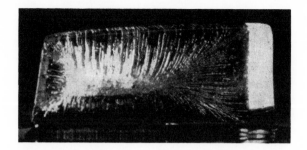

Fig. 6.7. Bubbles in ice cube. (From Ref. 41, p. 281.)

to as a "wormhole," can easily be seen in ice cubes produced by freezing water containing dissolved air; an example is shown in Fig. 6.7. It will be seen that the bubbles do not nucleate until the ice has grown inward for a few millimeters; this is presumably the distance required for the critical supersaturation to be produced. It is often observed that the "wormholes" in ice cubes oscillate in diameter, resembling a "string of pearls." This is attributed to periodic changes in the rate of solidification caused by the intermittent operation of the compressor

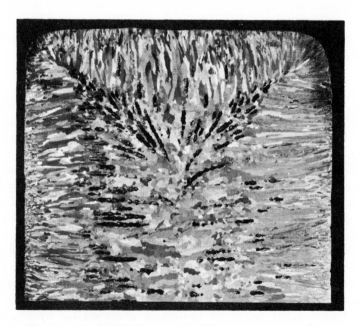

Fig. 6.8. Gas bubbles in a metal. (From Ref. 41, p. 281.)

of the refrigerator. As would be expected from the considerations discussed above, it is found that clear ice, that is, ice free from "wormholes," can be obtained by causing the water to flow continuously over the freezing interface. This prevents the concentration of dissolved air from reaching the critical level.

There have been no comparably detailed studies of the growth of gas bubbles in a solidifying metal, but there is ample evidence, of which Fig. 6.8 is an example, that the process is essentially the same as in water. It should be pointed out that the pressure at a point in a solidifying liquid may be lower than would be expected from purely hydrostatic considerations. In the extreme case, a region of liquid may be completely surrounded by solid. Further solidification, accompanied by decrease in volume, would rapidly decrease the pressure,

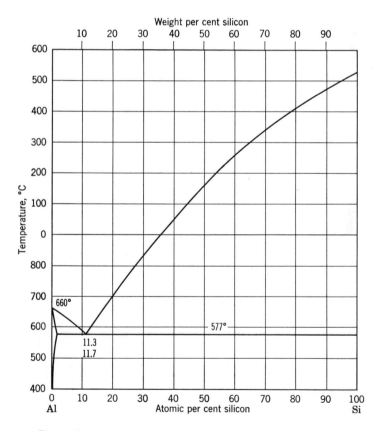

Fig. 6.9. Phase diagram for the aluminum silicon system. (From M. Hansen, *Constitution of Binary Alloys,* McGraw-Hill, New York, 2nd ed., 1958, p. 133.)

leading to the nucleation of bubbles. A decrease of pressure can also occur when the solidifying liquid connects to the bulk liquid only through narrow channels, such as those within and between cellular-dendritic crystals, where the hydrodynamic resistance to flow can cause a substantial pressure drop. The development of pores in a solidifying metal is discussed further in Chapter 8.

Formation of compounds by dissolved gases. A second possible effect of dissolved gases on solidification is that compounds, such as oxides, may be formed when the gas concentration reaches a high level, especially during the "terminal transient" stage of solidification. It is suggested by Tiller (3) that oxide and other inclusions that are often found at grain boundaries in cast metals may have been formed in this way, rather than being present in the liquid prior to solidification.

6.2 Eutectics

The intersection on a binary phase diagram of two liquidus lines that slope in opposite directions represents *eutectic equilibrium;* a

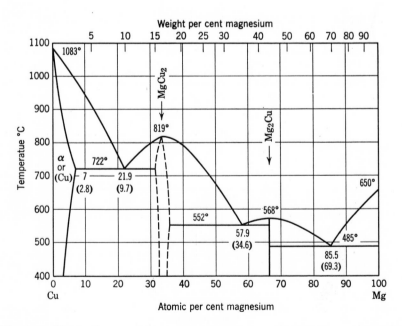

Fig. 6.10. Phase diagram for the copper magnesium alloy system. (From M. Hansen, *Constitution of Binary Alloys*, McGraw-Hill, New York, 2nd ed., 1958, p. 595.)

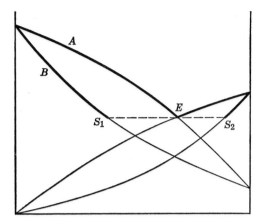

Fig. 6.11. Eutectic phase diagram.

single liquid phase is in equilibrium above the eutectic temperature, while 2 distinct solid phases constitute the equilibrium state below that temperature. All three phases can be in equilibrium at the eutectic temperature, but only at the eutectic composition. Examples of a eutectic point on a phase diagram are shown in Figs. 6.9 and 6.10. In each case, it can be seen that the eutectic point (or points) correspond to the point E in the schematic phase diagram of Fig. 6.11. The liquid at E is in equilibrium with S_1, by virtue of the relationship between the solidus A and the liquidus B, and with E_2 because of the other solidus-liquidus relationship. The essential condition for eutectic formation is that the solubilities should be limited, that is, that each species of atom should have a strong preference for its own crystal structure; or, when intermediate phases participate, these also must have limited solubilities for the neighboring phase.

The fact that the eutectic composition occurs at the interection of two downward sloping liquidus lines implies that it has the lowest liquidus temperature of any liquid in the part of the phase diagram between the neighboring maxima. This characteristic, namely that an alloy of two metals may have a lower melting point than either by itself, has long been known. The word "eutectic" is from a Greek word meaning "most fusible."

Most of the discussion of eutectic solidification will be based on binary eutectics, because ternary and more complex eutectic systems have not been subjected to sufficiently detailed study to elucidate their behavior, insofar as it differs from that of binary eutectics. A ternary eutectic lies at a minimum in the liquidus surface, and corre-

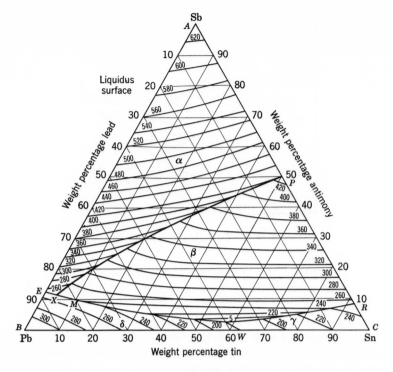

Fig. 6.12. The liquidus surface of the Pb-Sb-Sn system (X is the ternary eutectic point; liquidus temperature, 239°C). (From *Metals Handbook,* American Society for Metals, Cleveland, 1948, p. 1267.)

sponds to the simultaneous equilibrium of the liquid with three distinct solid phases. An example of a ternary eutectic phase diagram is shown in Fig. 6.12 in which the liquidus surface is represented by isothermal "contours."

Microstructure of eutectics.* A bewildering variety of structures are encountered in the metallographic examination of binary eutectic alloys; the only common feature is that two phases can always be seen. The eutectic structures that are observed can be classified into lamellar (in which *both* phases are lamellar), rod type, in which one phase is rod-shaped, and is embedded in a continuous matrix of the other phase, and discontinuous, in which 1 phase consists of isolated crystals embedded in a matrix of the other phase. There are various modifications of these basic structures, sometimes caused

* For a more detailed discussion of this and other aspects of eutectic solidification, the reader is referred to a recent review article by G. A. Chadwick (4).

by the tendency of the interface between the two phases to conform to specific low index planes of one phase or the other. It is a recent discovery (5, 6) that many eutectics are lamellar with a very regular structure if the metals used are sufficiently pure, and that many of the other structures that are observed are degenerate forms of the lamellar structure caused by impurities. It is therefore appropriate to consider first the simplest case, which is that of pure binary eutectics.

Pure binary eutectics. It was first shown by Chilton and Winegard (5) that if high-purity (zone-refined) tin and lead are used, the eutectic solidifies with an extremely regular lamellar structure; an example is shown in Fig. 6.13.

It is typical that the lamellae are straight and of uniform thickness except where an "offset" or "termination" occurs. It has been demonstrated by Chadwick (6) and by Crafts and Allbright (7) that both the uniform width of the lamellae and the presence of offsets are characteristic of pure eutectics. Chadwick has shown, for example, that in the aluminum copper eutectic the same lamellar structure can persist for several centimeters. It has also been established (7) that in a purely lamellar region of a eutectic, that is, where the lamellae are straight and parallel, each component has a nearly constant orientation, and that the orientations of the two components often have a simple crystallographic relationship to each other. Several investigations have been made on the orientation relationships, and although there are minor discrepancies, it is clear that typical relationships are

$$(100) \text{ Sn} \parallel (0001) \text{ Zn}; [001] \text{ Sn} \parallel [01\bar{1}0] \text{ Zn} \qquad (8)$$

$$(101) \text{ Sn} \parallel (111) \text{ Pb}; [010] \text{ Sn} \parallel [112] \text{ Pb} \qquad (9)$$

$$\text{Ag Cu: all planes and directions parallel} \qquad (10)$$

The example that has been investigated most thoroughly is that of the aluminum-copper eutectic, in which the two components are Cu and $CuAl_2$; Craft (11) has shown that the preferred relationship develops during the growth of a eutectic "single crystal"; this is a relationship not only between the orientation of the two phases, but also between the interlamellar surface and the crystal orientations. The relationship is expressed as

$$(111) \text{ Al} \parallel (211) \text{ CuAl}_2$$

$$[101] \text{ Al} \parallel (120) \text{ CuAl}_2$$

$$\text{Lamellar interface parallel to } (111) \text{ Al}$$

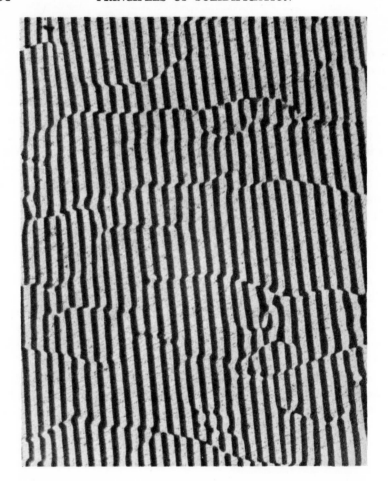

Fig. 6.13. Morphology of a pure eutectic (tin-lead). (From Ref. 4.)

It is found that these relationships are not precise, but have a tolerance of a few degrees.

Solidification of lamellar eutectics. Since it appears that the lamellar type of morphology is normal for eutectic solidification, it is desirable to discuss in general terms how a lamellar structure can form as a result of the solidification process. The first attempt to account for the existence of the lamellar structure was that of Tammann (12), who suggested that the layers of the two solid phases were laid down alternately, their interface being parallel to the solid-liquid interface. Tammann pointed out that while one phase is forming, the adjacent liquid is enriched in the other component, and suggested that when this enrichment reaches a critical level, the other phase forms as a layer over the previous one. This would give an alternation of layers of the two phases. Vogel (13), on the other hand, thought that the two phases grew simultaneously, so that their common interfaces would be perpendicular to the solid-liquid interface. It was shown experimentally by Straumanis and Braaks (14, 15) and confirmed by Winegard et al. (16) that the interlamellar interfaces were approximately normal to the mean solid liquid interface.

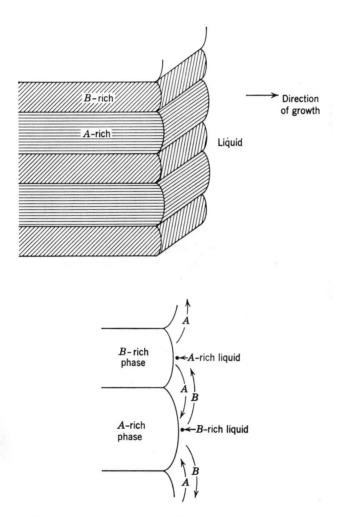

Fig. 6.14. Growth, mechanism, and diffusion paths for lamellar eutectic.

The more recent work on "pure" eutectics has fully confirmed the conclusion that both phases grow simultaneously, and that each lamella has its own solid-liquid interface. It follows that the liquid in front of each lamella becomes enriched in the major component of the neighboring lamellae, as shown in Fig. 6.14, and that transverse diffusion of both components must take place. If the liquid is exactly at the eutectic composition, it is clearly possible for the two components to be incorporated into the solid in exactly the same proportions that they have in the liquid; there will in this case be no long-range diffusion of either component forward into the liquid; the transverse diffusion should be substantially confined to a region comparable in thickness to the width of the lamellae.

It should be pointed out that a lamellar structure, or some modification of it, is the only one in which both of the phases can be continuous; a complete crystal consisting of many layers of each phase can originate from a single nucleus of each phase; presumably, a nucleus of one phase forms first, and as this grows the composition of the adjacent liquid changes until the conditions are reached at which the second phase nucleates on the surface of the first, which acts as a nucleant and may impose an orientation relationship. Further nucleation is unnecessary, as the process described by Winegard et al. (16) and called "bridging" by Tiller (3) can allow the growth of an existing lamella around the edges of lamellae of the other phase, giving rise to the alternation of the two phases.

Shape of the interface. The shape of the interface has been considered by Tiller ((3), who concludes that the shapes shown in Fig. 6.15 are possible. Figure 6.15a represents a shape that is determined by the approach to equilibrium between the interphase interface OA

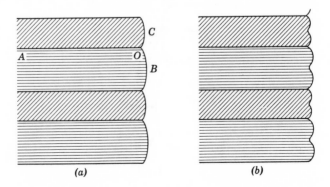

Fig. 6.15. Possible interface shapes.

and the two solid-liquid interfaces OB and OC in the immediate vicinity of the point (or, in the real three-dimensional case, line) O. If the use of the equilibrium concept is objected to in a system which is clearly not in equilibrium, it can instead be argued that since the interphase boundary has some free energy associated with it, the solid in its immediate vicinity is in equilibrium with the liquid at a lower temperature than is the case for the crystals themselves, and so the point O occurs at a position where the temperature is lower, that is, further to the left in Fig. 6.15. If the whole of the interface departs from local equilibrium to the same extent (which is the supercooling required for the kinetic driving force) then the curvature of the interface must everywhere (except at O) be such as to "match" the liquidus temperature of the liquid in contact with the interface. If it is assumed that the composition of the liquid in contact with the interface is constant over each lamella, then the resulting shape is that shown in Fig. 6.15a. If, however, the more realistic assumption is made that the composition of the liquid at the interface changes from a maximum concentration of solute at the center to a minimum at the edge of the lamella, then the profile sketched in Fig. 6.15b is to be expected. A more detailed examination of the shape of the interface has been made by Chadwick, Jackson, and Klugert (17), using an electrical analog method. Their results confirm that the shape normally has the convex form shown in Fig. 6.15a, and that in extreme cases it has the "concave" shape of Fig. 6.15b.

Lamellar growth: theoretical. There have been two distinct theoretical approaches to the relationship between lamellar width and rate of growth; it is assumed that the rate of advance of the interface is imposed by the conditions of growth, and that the widths of the lamellae adjust themselves to steady state values. Tiller (3), following a treatment by Zener (18) for a different but nearly analogous case, shows that the lamellar width is determined by the interaction of two opposing free energy considerations; one is that the transverse diffusion of the solutes must occur over greater distances for thicker lamellae, and therefore a larger driving force is required; and the other is that the free energy associated with the presence of the interphase boundaries increases as the lamellar width decreases; thus the diffusion condition would tend toward thin lamellae, while the surface energy of the interfaces favors thick lamellae. The amount of transverse diffusion that is required per unit time increases as the speed of growth increases, for a given lamellar thickness; therefore higher speed, with less time for diffusion, will produce thinner lamellae.

Tiller considers this problem to resemble the problem of the velocity of growth of free dendrites (Chapter 4) in that the equations governing, in this case, diffusion, are not in themselves sufficient to specify which of a range of results will actually occur. He therefore invokes an arbitrary criterion (that of minimum entropy production) without specific justification, and reaches the conclusion that the lamellar width d should be approximately proportional to $1/R^{1/2}$. If experimental values are inserted into the expression derived by Tiller, reasonable values are obtained for the energy of the interface between the lamellae of the two phases; this, however, requires the assumption of the value of a "shape factor" that takes into account the unknown geometrical conditions that govern the diffusion path between adjacent lamellae.

An alternative approach, that avoids the necessity for introducing an arbitrary criterion, has been developed by Jackson and Chalmers (19) who proposed the following physical basis for the steady state lamellar width. Figure 6.16 shows a feature that appears to be common to all lamellar eutectics; it is described as an "offset" or "termination." A mechanism by which a small change of lamellar width could occur as a result of a small change in speed of growth is for the terminating layer T to increase or decrease in length as growth proceeds. The locus of T, as growth proceeds in a direction normal to the plane of the diagram, would slope either to the right or to the left. This would cause a progressive change in the average lamellar spacing. The stability of the tip T is, therefore, the criterion for the stable lamellar width. The treatment proposed by Jackson and Chalmers is as follows: it is assumed that the interface is isothermal, and that

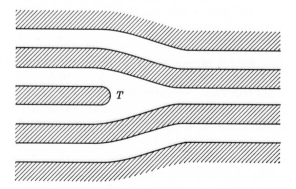

Fig. 6.16.   Termination of lamellar (schematic).

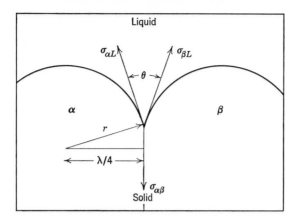

Fig. 6.17. Region of interface near the junction of two lamellae.

the total supercooling of the interface is the sum of the supercooling due to curvature and that resulting from the enrichment of the liquid in contact with the interface by rejection of the solute. The supercooling is calculated at the intersection of a termination with the interface, and at a position remote from terminations. The analysis has been developed only for the ideal case in which the widths of the two lamellae are equal and the curvatures are uniform and equal; that is, in which the surface free energies of the two phases (α and β) are equal. Consider a cross-section in the lamellar region of a eutectic, shown in Fig. 6.17. Resolving the surface energies in the vertical direction,

$$2\sigma_{\alpha L} \cos \frac{\theta}{2} = \sigma_{\alpha\beta}$$

also

$$\frac{\lambda}{4} = r \cos \frac{\theta}{2}$$

hence

$$\cos \frac{\theta}{2} = \frac{\sigma_{\alpha\beta}}{2\sigma_{\alpha L}} = \frac{\lambda}{4r}$$

or

$$\frac{\sigma_{\alpha L}}{r} = \frac{2\sigma_{\alpha\beta}}{\lambda}$$

The diffusion of solute ahead of the interface is given by $(C_\alpha{}^L - C_E) = [(1 - k_\alpha)C_E R\lambda]/8D$, where R is the rate and D the diffusion coefficient and therefore $\Delta T_c = [(1 - k_\alpha)C_E R \lambda m_\alpha]/8D$ where m is the slope of the liquidus line.

The supercooling at the center of a lamella associated with curvature is

$$\Delta T_r = \frac{\sigma_{\alpha L} T_E}{Lr}$$

but

$$\frac{\sigma_{\alpha L}}{r} = \frac{2\sigma_{\alpha\beta} T_E}{L\lambda}$$

and therefore

$$\Delta T_r = \frac{2\sigma_{\alpha\beta} T_E}{L\lambda}$$

At the termination point T (Fig. 6.16) $\Delta T_r = 4\sigma_{\alpha\beta} T_E/L\lambda$ because the surface at that point has spherical curvature instead of cylindrical curvature. The amount of solute rejected by the half cylinder of the termination (assumed to be stable) is $(1 - k_\alpha)C_E R (\pi/2)(\lambda^2/16)$ per unit time; this amount of solute diffuses across the semicircular interphase boundary, giving the following equation for diffusion:

$$(1 - k_\alpha)C_E R \frac{\pi}{2} \frac{\lambda^2}{16} = \frac{D(C_\alpha{}^L - C_E)\lambda/2}{\lambda/4} \pi \frac{\lambda}{4}$$

or

$$C_\alpha{}^L - C_E = \frac{(1 - k_\alpha)C_E R \lambda}{16D}$$

from which

$$\Delta T_c = \frac{m_\alpha(1 - k_\alpha)C_E R \lambda}{16D}$$

The sums of the two supercoolings are equated, giving

$$\lambda^2 R = \frac{32\sigma_{\alpha\beta} T_E D}{m_\alpha(1 - k_\alpha)C_E L}$$

from which $\lambda^2 R$ is a constant, or, $\lambda \propto R^{-\frac{1}{2}}$.

This solution is of the same form as Tiller's. However, it has the advantage of using a physical basis for the stable lamellar width instead of invoking the minimization of the rate of entropy production, the validity of which has been questioned (4).

Lamellar growth: experimental. Recent measurements by Chadwick (6) on Al–Zn, Al–Cu, Al–Zn, Pb–Sn, and Pb–Cd have shown that the relationship

$$\lambda \propto R^{-\frac{1}{2}}$$

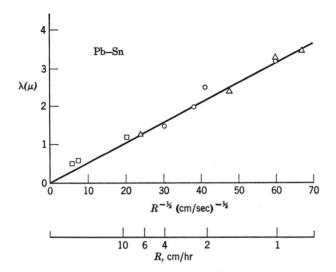

Fig. 6.18. Relationship between interlamellar spacing and growth rate for the lead-tin eutectic. (From Ref. 4.)

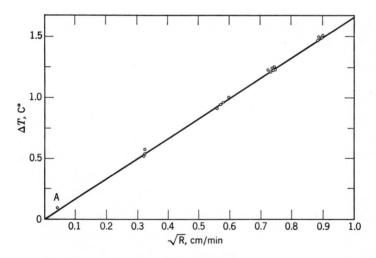

Fig. 6.19. Supercooling of eutectic interface as a function of growth rate (lead-tin). (From Ref. 4.)

is obeyed within experimental error, and that the temperature gradient in the liquid has no observable effect on the spacing; an example of Chadwick's results is given in Fig. 6.18. This result has been confirmed by Yue (20) for the system AlMg.

Hunt and Chilton (21) have recently measured the interface temperature of a growing lamellar eutectic and have found, as predicted by Tiller and by Jackson and Chalmers, that $\Delta T \propto R^{1/2}$, where ΔT is the supercooling at the interface with respect to the equilibrium temperature. The relationship is shown in Fig. 6.19.

Degenerate eutectic structures. While it is certain that the lamellar type of structure is characteristic of "pure" eutectics for a very wide range of solidification rate, Chadwick (6) has found that the structure degenerates at very slow rates of solidification (for example, less than 1 cm/hr). This degeneration, shown in Fig. 6.20 for $CuAl_2$ at 0.8 cm/hr, resembles the beginning of the spheroidization process that occurs during prolonged annealing. It has been demonstrated,

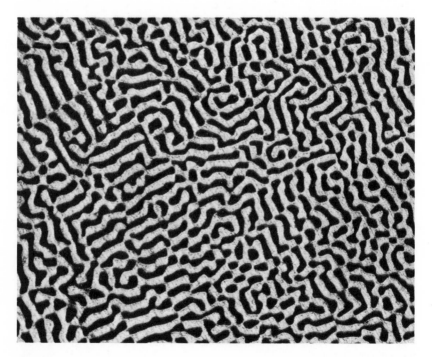

Fig. 6.20. Degenerate eutectic structure in $CuAl_2$-Al eutectic ($\times 500$). (From Ref. 4.)

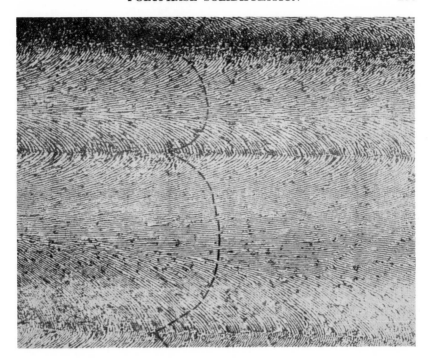

Fig. 6.21. Longitudinal section of impure CuAl₂-Al eutectic alloy. Broken line indicates shape of interface during growth. (From Ref. 4.)

however, that the degenerate structure is formed during, and not after, solidification.

Modification of eutectics. Two degenerate forms of the lamellar structure, both believed until recently to be alternative basic growth forms, are now known to be caused by impurities; these are the *colony structure* and the *rod structure;* they are described, following Chadwick, as modified structures.

COLONY STRUCTURE. It was proposed by Tiller (3) and demonstrated experimentally by Weart and Mack (22) that the colony structure of lamellar eutectics (Fig. 6.21) is associated with the presence of a cellular structure superimposed on the lamellar eutectic structure. The cellular structure is similar to the one described in Chapter 5, and results from the same processes. Tiller pointed out that an impurity, or an excess of one constituent, would diffuse much farther ahead of the interface than would be required for transverse interlamellar diffusion. This long-range diffusion sets up constitu-

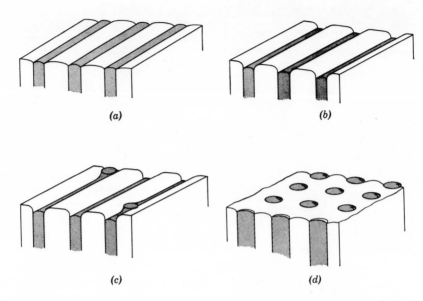

(a)   *(b)*

(c)   *(d)*

Fig. 6.22. Origin of "rod-type" eutectic structure (schematic). (From Ref. 4.)

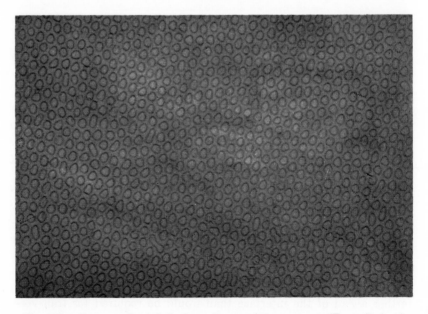

Fig. 6.23. Cross section of "rod-type" eutectic structure. (From Ref. 4.)

tional supercooling which is relieved by cell formation and the resulting transverse diffusion of the impurity. Chilton and Winegard (5) and Chadwick (6) showed that the cells and the colony structure are eliminated if the purity of the eutectic were sufficiently high, when the regular lamellar structure is produced.

Rod Structure. According to Chadwick (6), the lamellar structure is replaced by a rod structure when the impurity has sufficiently different distribution coefficients for the two solid phases; it will be realized that the colony, or cellular, structure merely requires a solute that has a distribution coefficient that differs from unity for both phases; for a pure colony structure they must be approximately equal. When the two distribution coefficients are very different, the lamellae of one phase should grow into the liquid ahead of the other, and the lamellae of the lagging phase then break up into very small cells, separated by the other phase. This is an alternative way of describing a rod type of structure. This is shown diagrammatically in Fig. 6.22 and a photomicrograph of a rod structure is shown in Fig. 6.23.

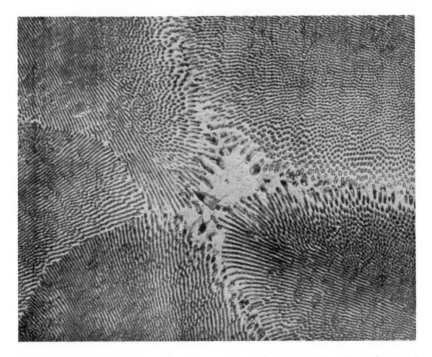

Fig. 6.24. Mixed lamellar and rod structure (Pb-Cd eutectic alloy with 0.1% Sn). (From Ref. 4.)

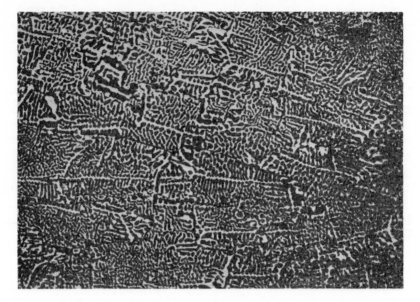

Fig. 6.25. "Chinese script" structure in Bi-Sn eutectic alloy.

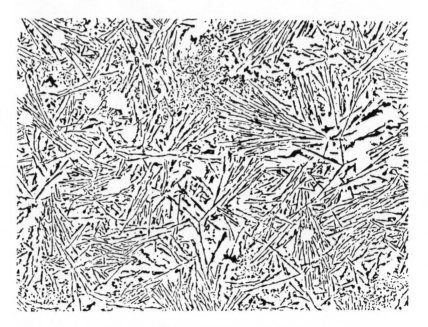

Fig. 6.26. Microstructure of Al-Si eutectic alloy.

An intermediate structure is that of colonies in which the middle is lamellar and the edges are rod-type. This is caused by an impurity which, when present at a low concentration, has nearly equal distribution coefficients for the two solid phases, but which has increasingly differing distribution coefficients as its concentration increases. The concentration is relatively low in front of the middle parts of the cells, and the similarity of the distribution coefficients leads to a lamellar structure. The concentration of impurity increases as the distance from the cell walls decreases; when the difference between the distribution coefficients is sufficient, the rod-type structure is formed. An example is shown in Fig. 6.24 for 0.1 per cent tin added to the lead-cadmium eutectic.

Discontinuous eutectics. In the lamellar type of eutectic and in its degenerate forms, each phase grows continuously, in the sense that continuous growth does not require repeated nucleation of either phase. This does not require that each phase is continuous with itself, although this is the case for a pure lamellar structure. A *discontinuous eutectic* is one in which one of the phases must renucleate repeatedly owing to the termination of growth of crystals of that phase. There appear to be two distinct types of discontinuous eutectic, typified by the tin bismuth eutectic (shown in Fig. 6.25) and the aluminum silicon eutectic (Fig. 6.26).

In each case, the discontinuity of the eutectic is apparently a result of a very specific morphology of the crystals of the discontinuous phase, which nucleate with random orientations and therefore grow in directions which are randomly oriented with respect to the growth interface. This leads to a situation shown schematically in Fig. 6.27 in which three

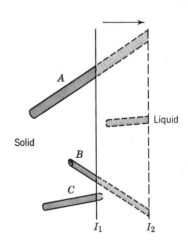

Fig. 6.27. Growth of a discontinuous eutectic (schematic), showing two positions of the interface (I_1 and I_2).

silicon crystals, A, B, and C, are shown, growing to the right as the interface I advances. The distance between A and B increases, with the result that the silicon concentration in that region eventually reaches a level at which nucleation of a new crystal (D) occurs. The crystal C, in the same way, stops growing because it is in compe-

Fig. 6.28. Spiral eutectic structure in Zn-Mg alloy. (From Ref. 23, p. 189.)

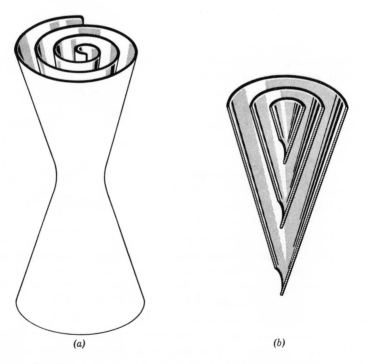

(a)   *(b)*

Fig. 6.29. Detailed structure of the spiral eutectic (schematic). (From Ref. 23, p. 191.)

tition with B for silicon. This is probably typical of all discontinuous eutectics, although the "Chinese Script" type has not been investigated sufficiently for the details to be clear.

The general explanation of discontinuity in eutectics is, therefore, the existence of strong anisotropy in the growth characteristics of one of the phases. If there is sufficiently strong growth anisotropy, the spiral type of discontinuous eutectic may be formed. The spiral structure has been found by Fullman and Wood (23) in the Al–Th and Zn–Mg eutectics. Figure 6.28 is a photomicrograph of a section of a spiral eutectic, and the actual structure is shown diagrammatically in Fig. 6.29.

The explanation proposed by Fullman and Wood for the formation of this unusual structure is that one or both of the phases is anisotropic in growth rate, in such a way that the alpha phase grows faster than the beta phase in one direction and more slowly in the other. The result is shown in Fig. 6.30; if the two edges of the beta phase do not form a closed ring, but overlap, then a spiral will be formed in that plane, and the complete structure will develop into a double conical spiral as shown in Fig. 6.29.

SPECIAL CASES OF THE MODIFICATION OF EUTECTICS. It has been known for a long time that the microstructure of the aluminum silicon eutectic could be modified by the addition of quite small amounts of the order of 0.01 per cent of, for example, sodium (24). The major effect is that the silicon phase, which in the unmodified alloy takes the form of needles or plates, occurs in very much smaller, more spherical particles in the modified alloy. A somewhat similar effect can be obtained by sufficiently rapid cooling, and it is found in both cases that the modified structure is formed at a temperature a few degrees below the normal eutectic temperature; however, the modified alloy melts at the same temperature as the normal eutectic. An explanation for these phenomena, proposed by Thall and Chalmers (25), is that the action of the modifier changes the surface tension relationships so that the aluminum phase, already "leading" the silicon by virtue of its lower latent heat and higher thermal conductivity, is able to "grow over," or "pinch off," the silicon crystals while the latter are still small. The effect of high speed of solidification would be to increase the importance of the thermal differences, and to reduce the size of the silicon crystals, so that "pinching" can occur without the help of the modifying agent. The lower temperature of solidification is caused by the necessity for constantly renucleating the silicon phase, for which some supercooling below the equilibrium temperature is required; this would occur when the silicon content in the liquid near

(a)

(b)

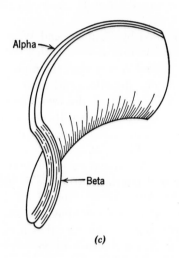

(c)

Fig. 6.30. Origin of spiral eutectic (schematic). (From Ref. 23, p. 192.)

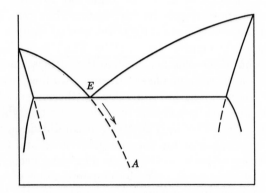

Fig. 6.31. Supercooling of eutectic in the absence of the second phase.

the interface increases as a result of the solidification of the aluminum rich phase alone, as shown in Fig. 6.31. The composition and temperature at the interface, in the absence of silicon crystals, would move from E toward A, corresponding to a continuous increase of supercooling below the silicon liquidus.

A more recent proposal by Rutter and Tiller (26) is that the distribution coefficient of the modifier is lower with respect to the silicon phase than to the aluminum phase; the growth of the silicon phase is therefore retarded because the liquid at its surface becomes progressively enriched with sodium; aluminum, already the leading phase, then grows across in front of the silicon in the constitutionally supercooled region resulting from the accumulation of sodium. This explanation appears to be adequate to account for the observations. Another theory (27), which has been widely accepted, is that the modifier "poisons" the nucleant particles so that more supercooling is required for nucleation than would be necessary for the pure eutectic. A serious objection to this theory, however, is that it fails to account for the fact that the silicon crystals grow only to a very small size in the modified alloy. In the unmodified alloy, nucleation of new crystals is rare, because once a crystal has nucleated it grows to a relatively large size; the temperature of the bulk of the solidifying liquid, therefore, remains close to the eutectic temperature, since a crystal of silicon that has nucleated quickly generates sufficient latent heat to restore the temperature to the steady state value. However, if the growth of the crystal is very limited, its latent heat would not raise the temperature appreciably and the mean interface temperature would be dominated by that of nucleation rather than by that of growth.

Noneutectic compositions. If the liquid is not initially at the eutectic composition, then the solidification process proceeds as follows. Let the initial composition be C_0 (Fig. 6.32). Solidification begins (neglecting supercooling) with the formation of solid of composition C_s; as solidification proceeds, the concentration of solute increases in the liquid close to the solid-liquid interface and this increase is distributed in the remaining liquid to a greater or lesser extent depending on the degree of mixing. The combination of rate of solidification, temperature gradient, and mixing determines the subsequent events. If mixing were complete, the whole of the liquid would change composition until it reached the point E, and crystals of cored composition, from C_s to C_T, would form. The remaining liquid would then solidify as a eutectic in the manner described above. The single phase solid that formed before the point E is reached would be described as *primary*,

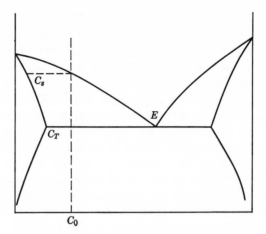

Fig. 6.32. Solidification of a eutectic system at a non-eutectic composition.

or *proeutectic* solid. If mixing is less complete, the primary solidification can take place dendritically, resulting in the cellular dendritic type of structure that is often observed; it is also possible for new crystals of the primary phase to be nucleated ahead of the interface as described on page 171. In any of these cases, the primary phase continues to grow until the remaining liquid has reached the eutectic composition. The solidification of the remaining liquid may then proceed by the normal eutectic mechanisms, forming eutectic in the interdendritic regions or in the regions between the grains of the primary structure.

It is important to observe that formation of eutectic in the last regions to solidify is to be expected even if the initial solidus is to the left of the point C_T in Fig. 6.32; this is the limit of composition at which eutectic would be expected *for equilibrium solidification*. However, it has been shown (page 136) that in real cases, the terminal transient liquid is far richer in solute than would be predicted from the equilibrium diagram, and it is therefore difficult to avoid the formation of some eutectic if the relevant liquidus line terminates at a eutectic point. An example is shown in Fig. 6.33.

Structure of eutectic liquids. Several investigations have provided evidence that some liquid alloys of eutectic composition have a tendency to form regions that are rich in one or another component. Vertman, Samarin, and Yakobson (28), for example, showed that centrifuging a liquid tin lead eutectic alloy caused separation that could be interpreted on the basis of "colonies" containing 10^3–10^4 atoms of lead. Lashko and Romanova (29), using x-ray diffraction, also found evidence for a quasi-eutectic structure in tin lead melts of eutectic composition (below 400°C); and in the Al–Ag system, similarly, evidence was found for the existence of the compound Ag–Al in the liquid; and AuSn was found in liquid gold-tin alloys of eutectic composition. Bartenev (30) showed that the evidence is consistent with the existence in the melt of regions of microsegregation of 25–50 A diameter, containing 10^3–10^4 atoms, probably with compositions close to the limit of solid solubility at the eutectic temperature.

This recent evidence is not inconsistent with the earlier work of Fisher and Phillips (31), for example, who demonstrated the existence of anomalies in the viscosity of some liquid alloys at the eutectic composition. However, it should be noted that the evidence for this anomaly is not conclusive; some investigators have failed to find any indication of it.

The evidence cited above seems to be in conflict with the views of Hume-Rothery and Anderson (32) who have analyzed the compositions of the eutectic points in binary metallic systems. They have

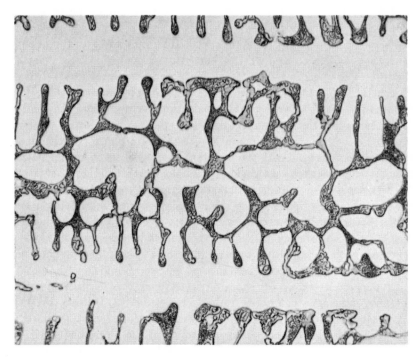

Fig. 6.33.   Interdendritic eutectic. (Photograph by H. Biloni.)

shown that the eutectic compositions are not randomly distributed in terms of atomic ratios, but that they occur preferentially close to certain specific compositions. Hume-Rothery and Anderson propose an explanation which depends upon the existence of rather sharp minima in the curves for free energy of the liquid as a function of composition. These minima would occur as a result of a type of local ordering in which a structural unit of the type proposed by Frank would contain both species of atoms in preferred sites. An objection to this theory is that eutectics are formed when "clustering," i.e., association of like atoms, is preferred energetically to "ordering" (association of unlike atoms). It is also open to doubt whether the liquid contains a sufficiently high concentration of these special structures for their atomic ratio to define the composition of the liquid as a whole.

Gravity segregation of eutectics. It has been shown by Allen and Isserow (33) that uranium aluminum alloys in the eutectic region of the phase diagram show marked segregation if they are "cycled" through the eutectic temperature. The segregation that is observed is an increase in the concentration of the uranium content in the lower part of the crucible, and, correspondingly, an increase in aluminum content at the top. The extent of the segregation increases as the number of cycles is increased. The effect is very large; e.g., a sample containing 13.3 per cent uranium cycled 168 times had a uranium concentration at the bottom of 45.4 per cent, as against 2.2 per cent at the top. The effect is attributed by Allen and Isserow to the "settling" of crystals of uranium (downward) and aluminum (upward); this explanation implies that separate nucleation occurs during the cooling cycle; however, the photomicrographs in the paper referred to show that the eutectic is lamellar, and the proeutectic phase is continuous, which is incompatible with the idea that there was sufficient nucleation to cause such marked segregation by the floating or sinking of crystals. An alternative explanation is that the segregation is in fact a result of the motion of the liquid enriched with solute during solidification and of the *purer* liquid formed by melting the separate phases during the melting part of the cycle. Since this effect is observed only when repeated cycles of melting and solidification take place, it does not throw further light on the structure of eutectic liquids.

Divorced eutectic. In some circumstances the terminal eutectic does not develop a characteristic eutectic structure; instead, the primary phase continues to solidify past the eutectic point (along the line EA) of Fig. 6.31 until either the whole of the liquid has solidified or the other phase nucleates and forms a layer, which is sometimes dendritic, separating the two layers of the primary phase. This type of terminal eutectic solidification cannot occur unless one of the phases requires considerable supercooling for nucleation. The term "divorced eutectic" is used to denote eutectic structures in which one phase is either absent or present in massive form.

Ternary eutectics. Very little work has been reported on the structure of ternary or higher eutectics. Winegard (34), however, has found recently that the ternary eutectic in the lead-tin-cadmium system solidifies in lamellar form, the three phases alternating as shown schematically in Fig. 6.34. This arrangement is the one which would provide the shortest possible diffusion path for a given total area of interphase boundary, since each phase is adjacent to both of the other two phases. No information appears to be available on quaternary or higher eutectics.

Cast iron. The term *cast iron* is used to represent any iron carbon alloy that has a carbon content between about 1.7 and 4.5 per cent; commercial cast irons always contain silicon, manganese, phosphorus, and sulfur, usually totalling at least two per cent, and it is necessary

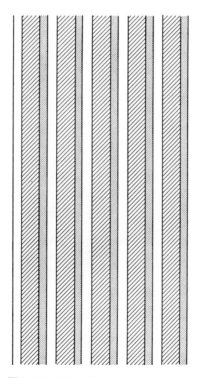

Fig. 6.34. Lamellar ternary eutectic.

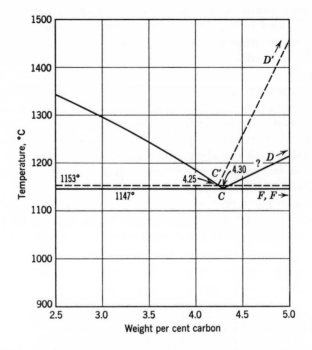

Fig. 6.35. Eutectic region of the iron carbon system. (From M. Hansen, *Constitution of Binary Alloys,* McGraw-Hill, New York, 2nd ed., 1958, p. 354.)

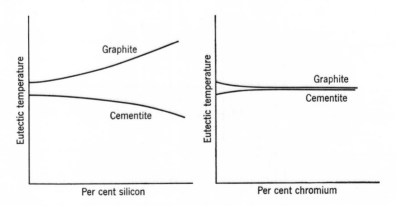

Fig. 6.36. Effect of third component on the eutectic temperatures (schematic).
(*a*) Silicon type, (*b*) chromium type.

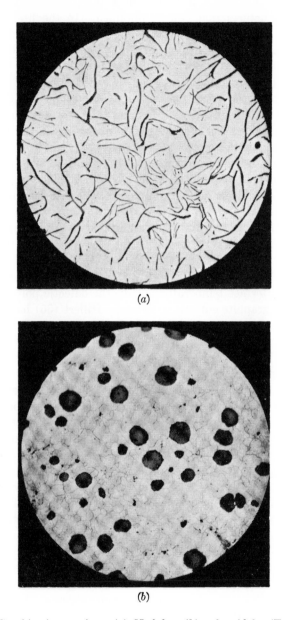

(a)

(b)

Fig. 6.37. Graphite in cast iron. (a) Nodular, (b) spheroidal. (From Ref. 41, p. 200.)

to take these constituents into account in discussing the very complex variety of structures that can be obtained. In the first place, it is known that two distinct eutectic systems can be formed; these are the iron-graphite and the iron-cementite systems. In the absence of the additional elements, the iron-graphite system is the stable one, but the iron cementite system can be formed by rapid cooling of the eutectic liquid, and both the graphitic and the iron carbide (Fe_3C) forms of carbon must be recognized as belonging to the system to be studied. The major classification of cast iron is according to the form of the carbon. If the carbon appears as graphite, the cast iron is described as gray; in white cast iron cementite is formed during solidification. The iron-cementite eutectic is called ledeburite. The term "mottled cast iron" is used when both graphite and cementite are formed during solidification.

The relevant part of the iron-carbon phase diagram is shown in Fig. 6.35 from which it is seen that the iron–Fe_3C eutectic temperature is about 6 degrees lower than that of iron-graphite. It has been shown by Morrogh and Williams (35) that if solidification takes place at an interface temperature above the eutectic point for cementite, then the graphite eutectic is formed. Hillert (36) has shown that a suitable nucleation agent is required. If, on the other hand, as a result of faster cooling, and, presumably, the formation of an enriched layer which depresses the liquidus, the interface is below the cementite eutectic temperature, then the white form of cast iron is obtained.

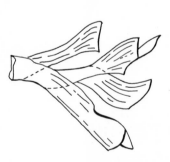

Fig. 6.38. Continuous graphite flake (schematic).

The effects of other elements can also be understood on this basis; silicon raises the graphite eutectic temperature and lowers the cementite eutectic temperature, as shown schematically in Fig. 6.36. It follows that the addition of silicon allows gray cast iron to be formed at higher solidification rates than would be possible without silicon. Chromium acts in the opposite direction, decreasing the temperature range in which graphite is formed. The morphology of the graphite is interesting; although typical photomicrographs suggest that the graphite exists as separate flakes (Fig. 6.37a), it has been demonstrated by Oldfield (37) that, when examined in three dimensions after dissolving out the iron, the graphite is continuous (Fig. 6.38).

SPHEROIDAL GRAPHITE. It has been shown that a process somewhat similar to the modification of the aluminum silicon eutectic can be used, with even more benefit to the mechanical properties, with gray cast iron. It was discovered by Morrogh and Williams (35) that the addition of small amounts of cerium to a cast iron of otherwise normal composition causes the graphite to form as discrete spherulets, instead of continuous flakes (Fig. 6.37b). The spherulets are interesting crystallographically; the orientation of the graphite is everywhere such that the basal plane of the structure (which is the low energy surface (38)) faces the melt. The high energy "edge" surface is nowhere exposed to the melt. The structure seems to be highly poly-

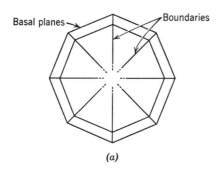

(a)

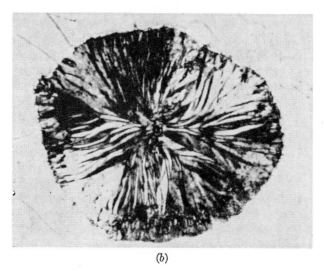

(b)

Fig. 6.39. Spherulet of graphite. (a) Schematic, (b) photomicrograph. (From Ref. 41.)

hedral, as sketched in Fig. 6.39a, with many low angle boundaries where the flat flake segments join; an actual spherulet is shown in Fig. 6.39b. It is not clear how or why the spherulitic form nucleates; it is probable, however, that it is the most stable form, energetically, in which the graphite can exist, since it combines a low surface area relative to its volume, because of its spherical shape, with the surface of lowest free energy per unit area. It is possible that the nucleus of the spherulet is formed by growth of graphite over the surface of a nucleant that is present as a small particle in the liquid. It has been shown by Morrogh (39) that a substantial fraction of the growth of the nodule occurs after it is surrounded by a layer of solid austenite; the carbon required for growth arrives by diffusion through the austenite. The stability of this form of graphite is confirmed by the fact that it also appears during long-term heat treatment of cast iron (malleableizing) in which it is to be expected that the most stable configuration will be approached. It has been demonstrated that one of the conditions required for the development of the spherulitic form is that the sulfur content of the melt should be very low; in fact it is possible that the cerium (or other agent such as magnesium, the effect of which was discovered subsequently) exerts its spheroidizing effect by combining with the sulfur, thereby taking it out of solution in the iron. The amount of spheroidizing addition required is proportional to the amount of sulfur present. In addition to the specific spheroidizing agent (Ce or Mg) we must, of course, add sufficient silicon (known as inoculant) to produce graphite rather than cementite.

6.3   Peritectic Solidification

A peritectic reaction occurs at the intersection of two liquidus lines that slope in the same direction; a schematic example is given in Fig. 6.40. At the peritectic point P, it is evident that the liquid is in equilibrium with the α phase because P is on the PQ liquidus line, and with the β phase because P is on the lower PR liquidus line. It follows that here, as in the case of the eutectic, one liquid is in equilibrium with two solid phases at a fixed composition and temperature; the difference is that only one of the solid phases can exist in equilibrium above the peritectic temperature, while in a eutectic system, each of the two solid phases can do so. It is usually stated that at the peritectic temperature the upper solid phase reacts with the liquid to form the lower solid phase. This is only possible under the unrealistic conditions of complete equilibrium; in this case, the liquid on cooling would form a mixture of α phase of composition S and liquid

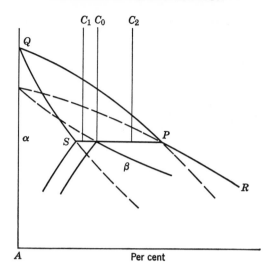

Fig. 6.40. Peritectic system, showing equilibrium phase boundaries ————, and nonequilibrium phase boundaries – – –.

of composition P when the peritectic temperature is reached. At this temperature (assuming complete equilibrium) the liquid and the α phase would react to form β; this reaction would go to completion. If the initial composition had been C_0, the whole of both phases would react, and the whole of the material would now be in the form of the β phase. If the composition had been C_1 or C_2, then the product would be β with an excess of α or of liquid respectively. The reaction $\alpha + L \rightarrow \beta$ is necessarily extremely slow, because except at the very beginning, the reacting phases, α and liquid, are separated by a layer of the product, β, through which diffusion must occur for the reaction to continue. As the layer of β thickens, the diffusion distance increases and the reaction becomes even slower.

The only investigation of what happens in real peritectic reactions is that of Uhlmann and Chadwick (40) who made qualitative theoretical predictions and verified them experimentally on the silver zinc system. Their argument shows that steady state solidification involving a planar two-phase (lamellar) interface, suggested by Chalmers (41), is impossible and that solidification occurs over a temperature range from the primary liquidus temperature to slightly below the peritectic temperature. A duplex structure is expected at all compositions between S and P of Fig. 6.40.

Consider a peritectic melt of composition M_1 (Fig. 6.41) and assume

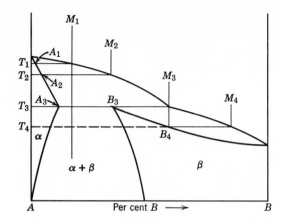

Fig. 6.41. Peritectic system. (From Ref. 40.)

that a linear temperature gradient moves through the liquid at a constant speed (Fig. 6.42); it will also be assumed that there is no convection in the liquid and no diffusion in the solid, i.e., that the solute moves by diffusion only. A fixed point in the liquid is considered; the events that occur there as the temperature gradient moves past

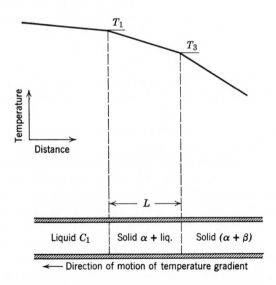

Fig. 6.42. Solidification of a peritectic in a temperature gradient. (From Ref. 40.)

will be as follows. When the temperature falls to T_1, primary crystals, of composition A_1, nucleate in the liquid. As the temperature falls to T_2, the crystals of α phase grow, or new ones nucleate. While this is taking place, the component B is rejected at the interface, and a layer of constitutionally supercooled liquid is formed. This layer of constitutional supercooling should result in the breakup of the assumed planar interface into the dendritic form, the length of the dendrites being determined by the thickness of the supercooled zone in the way described in Chapter 5. Part of the interdendritic liquid subsequently reaches the composition M_3, after which it solidifies as the β phase, finishing at a composition such as M_4. The predicted structure of the peritectic, therefore, is that of massive dendrites of the α phase, in a matrix of β phase. It might be added that if the temperature gradient were sufficiently steep, in relation to the growth rate, the dendritic structure should be replaced by a cellular structure in which the cell walls consist of the β phase. The predicted dendritic growth of the α phase was verified over a wide range of compositions (within the limits of the peritectic horizontal) and speed of growth, in experiments in which the direction and speed of solidification were controlled. It therefore appears certain that peritectics do not solidify by the lamellar mechanism characteristic of eutectics, but with a distinctive morphology not previously recognized.

6.4 Solidification in the Presence of a Solid Phase

Suspended particles. It has been assumed up to this point that the liquid has behaved as a single homogeneous phase, except for the possible effects of nucleant particles. However, most, if not all, liquid metals contain particles of solid in suspension; their distribution in the resulting solid may have important consequences, both by their influence on the dislocation content (see page 58) or directly on the mechanical properties. It is, therefore, relevant to consider the interaction between an advancing solid-liquid interface and solid particles in the liquid. The problem to be considered, therefore, is how the solidification process affects the distribution of particles in suspension in the liquid. There are three factors that may influence the final location of a particle: (1) If the density of the particle is different from that of the liquid, it will tend to float or to sink. The extent to which the behavior of the particle is dominated by its buoyancy (either positive or negative) depends upon the density difference and on the size and shape of the particle. A particle that is sufficiently small will remain in suspension indefinitely as a result of its Brownian motion even if its density is substantially different from that of the liquid. The actual size at which the Brownian motion becomes effective depends on the density difference, but in general is of the order of $\frac{1}{10}$ μ. For large particles, the rate of ascent or descent B is given for a sphere by the Stokes formula:

$$B = \frac{2}{9} \frac{gr^2(D_1 - D_2)}{\eta}$$

For nonspherical shapes, the value of B is smaller because a particle always tends to orient itself so that it offers the maximum resistance to its own motion through the liquid. For a particle of 1 μ radius, with a density difference of 2 gm/cm^3, the velocity would be of the order of 10^{-4} cm/sec.

(2) The second factor is the motion of the fluid, which can often be large enough to maintain in suspension particles that would sink or float in a stationary liquid. The motion of the fluid that is generated as the liquid enters the mold may persist for a considerable time before it gives way to convection caused by thermal and compositional gradients.

(3) It follows that although there may be some vertical separation due to flotation or sedimentation, and some radial separation resulting from centrifugal forces, the smaller particles may remain suspended with a nearly random distribution. The final distribution in the solid depends upon whether a particle is "trapped" *in situ* by the advancing solid-liquid interface or whether it is pushed ahead as the interface moves forward.

There appears to be no experimental information on the interaction between an advancing solid-liquid interface and suspended particles in metallic systems; however, Uhlmann and Chalmers (42) have studied this problem in some nonmetallic systems, whose transparency allows direct visual observation of the behavior of particles in the vicinity of an interface. Although the results of this work may not apply to the somewhat different case of metals, the general conclusions are probably relevant. It was found that for any given system (such as particles of MgO in orthoterphenyl) there is a critical velocity of advance of the interface, below which particles are pushed ahead of the interface. At higher speeds the particles are trapped in the positions they occupied when the interface arrived. In the particular case cited the critical speed is about 0.5 μ/sec; other combinations of matrix and particles gave critical velocities ranging from zero (that is, less than about 0.1 μ/sec) up to about 2.5 μ/sec. There is no obvious chemical or crystallographic criterion that correlates with the critical speed; it does not appear to depend as directly as might be expected on the potency of the particles as a nucleation catalyst for the matrix material, perhaps because in heterogeneous nucleation, the nucleus probably does not form unless a specific low-energy orientation relationship is satisfied between the crystal and the substrate, while a suspended particle is unlikely to be appropriately oriented with respect to the advancing interface to form a low-energy interface. A surprising feature of the experimental results is that, for a given system, the critical speed was found to be independent of the size of the particle over a wide range of sizes. It was estimated that the force exerted by the growing interface on the particle was of the order of 10^{-4} dynes for a 5-μ particle. This force can be regarded as arising from an increase in the surface energy when the crystal-liquid and liquid-particle interfaces are replaced by a single crystal-particle interface. The analysis of the data shows that the particle is not pushed significantly until it is almost within molecular dimensions of the interface, and that the critical speed arises from the problem of "feeding" liquid into the space between the particle and the growing crystal.

It should be pointed out that the critical speeds that were determined (up to 2.5 μ/sec., or about 1 cm/hr) are very slow compared with most practical solidification or crystal growing procedures. If the critical speeds for metals or semiconductors are equally low, therefore, there would be few if any cases in which the distribution of particles would be changed by the solidification process.

Solidification of a liquid in a porous solid. Little attention has been paid to the solidification of a liquid metal that is contained in interconnected channels in a porous solid that is chemically inert to the solidifying liquid; an equivalent problem in a nonmetallic system is encountered in the freezing of water in soil. In this case large forces, exerted by the freezing water, give rise to frost heaving, which causes serious damage to highways and other structures. These forces arise not because water expands on freezing, but because a water layer persists between ice and soil particles. As ice is formed, more water is drawn into the region of contact to replace what has frozen. This water in turn starts to freeze, causing more water to be "sucked" in, and forcing the existing ice away from the soil particles. This process is not unlike the pushing of solid particles in suspension, discussed on page 229; it depends upon the same surface energy relationship, that is, a preference, energetically, for the existence of a liquid layer between the two solids. A liquid metal contained in a porous matrix may have a similar surface energy relationship, in which case very large forces could be exerted, tending to disrupt the matrix.

References

1. M. Smialowski, *Hydrogen in Steel*, Addison Wesley Publ. Co., Reading, Mass., 1961.
2. B. Chalmers and J. Newkirk, Unpublished Work.
3. W. A. Tiller, *Liquid Metals and Solidification*, American Society for Metals, Cleveland, 1958, p. 276.
4. G. A. Chadwick, *Progr. in Materials Sci.*, **10**, 97 (1963).
5. J. P. Chilton and W. C. Winegard, *J. Inst. Met.*, **224**, 1176 (1961).
6. G. A. Chadwick, *J. Inst. Met.*, **91**, 169 (1693); **92**, 18 (1963).
7. R. W. Crafts and D. L. Allbright, *Trans. Met. Soc. AIME*, **224**, 1176 (1962).
8. W. Straumanis and N. Braaks, *Z. Physik. Chem.*, **38B**, 140 (1937).
9. N. Takahashi and K. J. Ashinuma, *J. Inst. Met.*, **87**, 19 (1958–1959).
10. E. C. Ellwood and K. Q. Bagley, *J. Inst. Met.*, **76**, 631 (1949).
11. R. W. Craft, *Trans. Met. Soc. AIME*, **224**, 65 (1962).
12. G. Tammann, *Textbook of Metallography*, Chemical Catalogue Co., New York, 1925, p. 182.
13. R. Z. Vogel, *Anorg. Chem.*, **76**, 425 (1912).
14. W. Straumanis and N. Braaks, *Z. Physik. Chem.*, **29**, 30 (1935).
15. W. Straumanis and N. Braaks, *Z. Physik. Chem.*, **37**, 38 (1937).
16. W. D. Winegard, S. Majka, B. M. Thall, and B. Chalmers, *Can. J. Chem.*, **29**, 320 (1957).
17. G. A. Chadwick, K. A. Jackson, and A. Klugert, Unpublished; referred to in Ref. 4.
18. C. Zener, *Trans. AIME*, **167**, 550 (1946).
19. K. A. Jackson and B. Chalmers, Unpublished, referred to in Ref. 4.
20. A. S. Yue, *Trans. Met. Soc. AIME*, **224**, 1010 (1962).
21. J. D. Hunt and J. P. Chilton, *J. Inst. Met.*, **92**, 2214 (1963–1964).
22. H. Weart and J. D. Mack, *Trans. AIME*, **212**, 664 (1958).
23. R. L. Fullman and D. L. Wood, *Acta Met.*, **2**, 188 (1954).
24. Z. Jeffries, *Chem. Metall. Eng.*, **26**, 750 (1962).
25. B. M. Thall and B. Chalmers, *J. Inst. Met.*, **77**, 79 (1950).
26. W. A. Tiller, *Acta Met.*, **5**, 56 (1957).
27. R. C. Plumb and J. E. Lewis, *J. Inst. Met.*, **86**, 393 (1957).
28. A. A. Vertman, A. M. Samarin, and A. M. Yakobson, *Izvest. Akad. Nauk. SSSR Met. i Topl.* (Tekhn), **17** (1960).
29. A. S. Lashko and A. V. Romanova, *Izv. Akad. Nauk. SSSR Met. i Topl.*, (Tekhn) 135 (1961).
30. G. M. Bartenev, *Izv. Akad. Nauk. SSSR Met. i Topl.* (Tekhn) 138 (1961).
31. H. Fisher and A. Phillips, *Trans. AIME*, **200**, 1060 (1954).
32. W. Hume-Rothery and E. Anderson, *Phil. Mag.*, **5**, 383 (1960).
33. B. C. Allen and T. Isserow, *Acta Met.*, **5**, 465 (1957).
34. W. C. Winegard, *Private Communication*.
35. H. J. Morrogh and W. J. Williams, *J.I.S.I.*, **155**, 321 (1947).
36. M. Hillert, *Trans. ASM*, **53**, 555 (1961).
37. W. Oldfield, *BCIRA J.*, **8** (1960).

38. W. Patterson and D. Amman, *Giesserie,* **23,** 1247 (1959).
39. H. J. Morrogh, *J.I.S.I.,* **176,** 378 (1954).
40. D. Uhlmann and G. A. Chadwick, *Acta Met.,* **9,** 835 (1961).
41. B. Chalmers, *Physical Metallurgy,* John Wiley & Sons, New York, 1959, p. 272.
42. D. Uhlmann and B. Chalmers, Unpublished Work.

7

Macroscopic Heat Flow
and Fluid Flow

7.1 General Considerations

Any process that is based upon the solidification of a metal has the
purpose of producing a piece of solid metal that fulfills certain requirements, the exact nature of which depends upon the application;
however, it is almost always true that the requirements can be divided
into two kinds, namely, the geometrical and the structural. The geometrical considerations are, broadly, that the external shape must be
satisfactory, and that internal voids, if they exist, must be within
permissible limits of size, shape, and location. If the solidified metal
is acceptable on the basis of its geometrical characteristics, then the
question may arise as to whether its properties are adequate, and this
is determined by its structure. Before considering in detail the
interaction of the various factors that control the structure and the
geometry, however, it is necessary to review the problems associated
with the flow of metal into a mold and the extraction of heat from
the metal. These two problems are by no means independent of each
other, because loss of heat by the metal while it is flowing into a mold
is often a limiting process.

7.2 Fluid Flow

The ability of a molten metal to flow has two important implications; the first is that it can be poured from a container in which it
was melted into a mold in which it is to solidify. The second is that
relative motion of different parts of the liquid can occur while it is
solidifying. The former may affect the macroscopic geometry of the
casting, and will be considered here, while the latter, on account of
its implications in relation to the structure of the solidified metal, will
be discussed in Chapter 8.

Viscosity of liquid metal. Liquid metals, in common with other liquids, are viscous, in the sense that the rate of flow depends upon the force, or, to be more precise, the rate of shear is proportional to the shear stress. This means that the rate of flow of a liquid through a tube, for example, depends upon the pressure difference between the ends of the tube, on its length, and on the radius of the tube. The relevant formula is

$$Q = \frac{\pi r^4}{8\mu} \cdot \frac{P_1 - P_2}{l}$$

When Q is the quantity flowing per unit time, r is the radius of the tube, P_1 and P_2 are the pressures at the two ends of the length l, and μ is the viscosity.

The formula given above applies only in cases in which the flow is of the "stream-line" or laminar type, which occurs at relatively slow rates of flow. To compare "rates of flow," in this sense, it is necessary to take into account the viscosity μ, density γ, velocity v, and the linear dimension l, of the system; the behavior will be similar if $\gamma v l/\mu$ is the same for two systems; this quantity is known as *Reynolds' number*. If the value of Reynolds' number is high, for example, greater than 1400 for a tube leading out of a containing vessel, the flow becomes turbulent and Q drops below the value that would be calculated from the above formula. It will be noticed that the ratio of viscosity to density, μ/γ, occurs in the expression for Reynolds' number; the same ratio enters into the expression for the rate of flow through a tube if the pressure difference is caused by a head of the liquid that is flowing. The quantity μ/γ is called the *kinematic viscosity*, and is much more significant for our present purpose than the viscosity μ. The values of viscosity and kinematic viscosity of some liquid metals are given in Table 7.1; the values are also given for water for comparison. The best available evidence (1, 2) indicates that the viscosity

Table 7.1

Metal	Viscosity (poise)	Kinematic Viscosity (cm²/sec)
Mercury	0.021	0.0012
Lead	0.028	0.0025
Tin	0.020	0.00231
Copper	0.038	0.0047
Iron	0.040	0.0050
Water	0.010	0.010

changes rather slowly with temperature, and that this applies even below the melting point into the region of supercooling; however, there have been reports (3, 4, 5, 6) that the viscosity rises rapidly as the temperature falls toward the melting point. It is difficult to reconcile these results with the fact that metals still exhibit characteristic liquid properties when supercooled, and it is thought that the rapid increase is a spurious experimental effect (7). It may be concluded from the foregoing discussion that liquid metals, when they are *completely* liquid, flow rather more easily than water, and that their viscosity is seldom, if ever, a limiting factor in the process of filling a mold, even through a rather narrow channel.

There is, however, a hydrodynamic effect that can be important; this is expressed by the well-known theorem of Bernoulli, which states that, for a flowing liquid,

$$\frac{p}{w} + Z + \frac{q^2}{2g} = \text{constant}$$

where p/w is the pressure due to head of liquid, Z is the height above some arbitrary level, and q is the velocity of flow. A consequence of this is that if the velocity of flow q of a liquid increases, for example, because of a constriction in the tube through which it is passing, then the pressure p of the liquid must decrease. A liquid metal flowing through a complicated gating system may therefore decrease in pressure to such an extent that bubbles of air are sucked through the mold wall and entrained in the flowing liquid. The entrained air may remain in the liquid until it is in the mold, where it can cause internal voids in the casting.

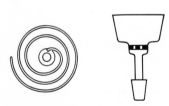

Fluidity. When a mold is to be filled through a gating system, or when the mold has thin sections, it is often found that the metal fails to fill the mold completely, owing to solidification in narrow channels in the gating system or in the mold. This difficulty is attributed to

Fig. 7.1. Mold for fluidity test. (From *Metals Handbook,* American Society for Metals, Cleveland, 1948, p. 200.)

lack of *fluidity,* a term that must be distinguished clearly from *viscosity.* Fluidity is defined (8) as "the ability of a liquid metal to flow readily, as measured by the length of a standard spiral casting." There are a number of "standard" fluidity molds, of which one of the simplest is shown in Fig. 7.1. The fluidity is defined as the length of the casting from the point where the metal enters, in this case at the center, to the limit which it reaches. The measured fluidity is very sensitive to the temperature at which the metal is poured (i.e., to the amount of superheat), and for alloys it is to some extent a measure of the solidus-liquidus interval. The fluidity is actually a measure of the way in which solidification proceeds inward from the wall of the spiral mold as the metal flows along it, and it therefore depends upon the solidification characteristics (including nucleants) rather than on the ability of the completely liquid metal to flow. It is not clear that the fluidity as measured in this way correlates significantly with the actual behavior of alloys in the foundry (9).

7.3   Heat Flow

It has been shown (Chapter 2) that the rate at which solidification takes place at any point of a solid-liquid interface depends upon, and is roughly proportional to, the departure of the actual temperature from the equilibrium temperature. In order to maintain a temperature difference, however, it is necessary to remove the latent heat of fusion at a rate equal to that at which it is generated; and it is therefore valid, from a macroscopic point of view, to consider that the amount of solidification that takes place in a given time is controlled by the amount of heat removed in that time. It is possible, in principle, to calculate the rate at which heat would be removed from a given mass of liquid metal, at a known initial temperature, after it has been, for example, poured into a mold at some lower temperature. It would be necessary to know the appropriate physical constants for the metal and the mold and for heat transfer between them; but the total amount solidified could be calculated as a function of time.

The heat-flow problem is intractable in its most general form, in which the solid-liquid interface temperature changes with time owing to the accumulation of rejected solute, and if the mold is of finite thickness and has a prescribed realistic geometry. However, as will be shown, certain simpler cases have yielded to analysis, and it is also possible to develop quantitative information by the Analog Computer method.

From the point of view of heat extraction, and therefore of the rate of solidification, it is useful to classify the various solidification processes in terms of the thermal properties of the mold. At one extreme, that of a metal mold, the thermal conductivity of the mold is comparable with that of the solidifying metal; the relevant thermal property is the *heat diffusivity* b defined as $\sqrt{K\gamma C}$, where K is the thermal conductivity, γ is the density, and C is the specific heat. The ratio n of the values of b for the mold material and the solidifying metal is the criterion that will be used. Horvay and Henzel (10) give the value $n = 1.12$ for steel in a cast iron mold; some relevant properties are given in Table 7.2. If the thickness of the mold is not

Table 7.2   (Values in feet, pounds, °F units)

Material	K	γ	C	$b = \sqrt{K\gamma C}$
Aluminum	120	170	0.26	73
Copper	224	560	0.10	112
Steel (solid)	18.4	460	0.16	37
Cast iron	20	460	0.15	37
Sand	0.90	94	0.28	3.6
Graphite, 1500°	19	140	0.29	28
1000°	67	140	0.29	52

large compared with the thickness of the section that is to be cast, then, as will be shown, the early part of the solidification process is controlled by heat flow into the mold, while the later part depends upon conduction through the mold and heat loss from the outside. Three types of metal "molds" must be distinguished: (a) A metal mold that loses heat from its outer surface only by radiation and convection; this is typical of ingot molds; (b) a metal mold that is cooled on the surface remote from the casting. Most continuous castings are made in molds which are water cooled; water cooling of the part of the casting that has emerged from the mold has a similar effect (see discussion of continuous casting). The molds that are used for arc melting are also of this type; and (c) a metal "mold" that is massive compared with the solidifying metal, so that heat loss from the outside of the mold does not become significant until after solidification is complete. The extreme example of this is a weld (Fig. 7.2), in which a bead or pool of molten metal, often with substantial superheat, is

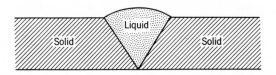

Fig. 7.2.   Cross section of butt weld (schematic).

in good thermal contact with the "work," into which heat is rapidly conducted.

The second type of mold-metal relationship is typified by the sand mold, whose thermal conductivity is so much lower than that of the solidifying metal that temperature gradients in the latter can usually be neglected. The value of n in this case is given by Horvay and Henzel as 0.13. Here the thickness of the mold is much less critical, because the heat loss from the outer surface of the mold is comparable with the effect of extra mold thickness.

The third type of thermal control of solidification is that in which heat extraction is limited by controlling the supply of heat to the liquid metal, as in zone-refining and most methods for growing single crystals, and by limiting the heat extraction by means of a controlled heat sink; the rate of the process is usually determined by the rate of motion relative to the sample of a controlled temperature gradient.

Rate of solidification. Much of the discussion of the heat-flow problem in the literature is focussed on the question of how the thickness of the "solid layer" in contact with the mold increases with time. This question is significant only in the special case in which the metal solidifies with a planar interface, and this is likely to arise under practical casting conditions only with a pure metal or a eutectic alloy, in which the freezing range (i.e., interval between solidus and liquidus) is zero. In other alloys, the interface is almost always dendritic or cellular-dendritic, if indeed there is a coherent interface at all. The factors that determine the type of interface are discussed in Chapter 8. The interpretation of the results of the heat-flow calculations is, therefore, not as simple as has sometimes been assumed; comparison with experiment is not easy because, as pointed out by Ruddle (11) the bulk of experimental evidence is based on the "pour out" method, which may indicate too much metal frozen, if liquid is trapped in interdendritic spaces, or too little, if there are "independent" crystals floating in the liquid. A much more satisfactory method of experiment is the measurement of the temperature-time relationship for various points within the solidifying metal. This has provided most of the useful experimental information on the rate of solidification of a metal in a mold; a detailed account of the techniques that have been developed is given by Ruddle (11).

In spite of the limitations of the analytical approach, it is worth while to consider some of the results that have been achieved. The simplest case to consider is that in which the metal and the mold are both semi-infinite, the metal-mold interface temperature remains constant, and the metal is initially at the melting point (i.e., no superheat). It is assumed, moreover, that the liquid is a pure metal, so that the solid-liquid interface temperature is constant.

The general method of approach is as follows:[*]

For heat conduction in one direction in the mold (i.e., perpendicular to the planar mold wall), the temperature θ of an element of volume at time t is given by

$$\frac{\partial \theta}{\partial t} = \frac{K}{\gamma C} \frac{\partial^2 \theta}{\partial x^2}$$

where K = thermal conductivity
γ = density
and C = specific heat

If a plane-bounding surface of a semi-infinite solid, initially at θ_0, is instantaneously raised to θ_i at time $t = 0$, then after time t_1 the temperature θ_m at any point is given by

$$\theta_m = \theta_0 + (\theta_i - \theta_0) \, \text{erfc}\left(\frac{X}{2\sqrt{\alpha t_1}}\right)$$

where

$$\alpha = \frac{K}{\gamma C}$$

The rate $\partial Q/\partial t$, at which heat is removed from the casting at any time t_1, is given by

$$\frac{\partial Q}{\partial t} = -K \left[\frac{\partial \theta}{\partial x}\right]_{x=0}$$

which is equivalent (by differentiation of the erfc equation) to

$$\frac{\partial Q}{\partial t} = \frac{K(\theta_i - \theta_0)}{\sqrt{\pi \alpha t}} = 0.564 \frac{K(\theta_i - \theta_0)}{\sqrt{\alpha t}}$$

$$= 0.564b \frac{(\theta_i - \theta_0)}{\sqrt{t}}$$

where b is the heat diffusivity $\sqrt{K\gamma C}$.

The total heat conducted into the mold up to the time t is obtained by integrating the previous equation:

$$Q = b(\theta_i - \theta_0) \int_0^t \frac{0.564}{\sqrt{t}}$$

$$= 1.128b(\theta_i - \theta_0)\sqrt{t}$$

[*] This discussion is essentially that of Ruddle (11).

This may be rewritten as

$$D = q\sqrt{t}$$

where D is the thickness of the solidified layer, and q, the solidification constant, is 1.128 $[b(\theta_i - \theta_0)/L\gamma']$; L and γ' are the latent heat of fusion and density of the solidified metal.

It will be appreciated that the equation $D = q\sqrt{t}$ does not accurately represent any real case; it is assumed that the mold wall is so thick that all the heat conducted from the metal is used to heat up the mold, and that there is still some mold material which is not affected; it is also assumed that the thermal conductivity of the metal is so high compared with that of the mold that the temperature gradient required for conduction of heat through the metal can be ignored. Therefore the solid-metal mold interface remains at the same temperature as the solid-liquid interface; however, the essential conclusion that the rate of thickening of the frozen "skin" is proportional to the square root of time is a useful approximation.

If superheat is to be taken into account, the problem becomes more complicated because (1) the extra amount of heat to be conducted into the mold depends upon the volume of metal in the mold, which can no longer be regarded as infinite, and (2) the initial temperature of the metal-mold interface is not the same as the solid-liquid interface temperature which, instead, changes during solidification. However, a good approximation can be obtained by adding in the superheat in the form: total heat to be extracted from solidifying metal is equal to

$$W[L + S(\theta_c - \theta_f)]$$

where W = weight of casting
     θ_c = initial temperature of the liquid
     θ_f = final solidification temperature

This amount of heat must be conducted into the mold during the time t taken for solidification. If the area of the surface of contact of mold and metal is A, then

$$\sqrt{t} = \frac{W[L + S(\theta_c - \theta_f)]}{1.128A\sqrt{K\gamma C}\,(\theta_i - \theta_0)}$$

from which

$$t = \left(\frac{V}{A}\right)^2 \times \text{constant}$$

so long as the pouring temperature and the other material properties remain constant. This is known as Chvorinov's rule. It should be noted that this part of the discussion applies to the time for complete solidification, and still assumes that all heat flow is perpendicular to the plane of the mold surface, a condition that could be met in practice only by a mold whose lateral extent is large compared to the thickness of the solidifying metal. It is probable that in real cases, at least in a sand mold, the effect of superheat is to delay the start of solidification rather than to extend the time during which it occurs. This is because the high conductivity and convective motion of the liquid in the mold causes the whole of the metal to cool at essentially the same rate, so that it is all very close to the liquidus temperature before any solidification begins. Ruddle and Mincher (12) found that, for copper, aluminum, magnesium and zinc, and for a copper-aluminum alloy close to the eutectic composition, cast into a sand mold with a superheat of 100°, the results are well represented by

$$D = q\sqrt{t} - c$$

A somewhat different approach to the same problem is that of Horvay and Henzel (10) who present nomograms for the solidification parameter ω which is $q/2\sqrt{b}$, where q is the solidification constant (as before) and b is the temperature diffusivity, $K/\gamma C$. Horvay and Henzel compare the results of these calculations with the experimental results of Pellini and his coworkers (13), and they point out that a major source of discrepancy is their neglect of the effect of the air gap that forms between the outside of the casting and the inside of the mold, soon after the solid skin is formed, as a result of the thermal contraction of the casting as it begins to cool and the thermal expansion of the mold as its temperature rises.

Horvay and Henzel also show how to take into account the effect of the finite thickness of the layer of liquid metal, using the "reflection" of the temperature "wave" at the bounding surface. The solidification rate for a planar interface slows down continuously, according to the parabolic law; however, as shown by Adams (14) for the more realistic cylindrical shape, in which all the heat flow is radial, the solidification rate reaches a minimum and thereafter increases, because the area of the solid-liquid interface, and therefore the latent heat per unit distance of advance of the interface, decreases as the interface approaches the center. The minimum linear rate of solidification occurs after freezing has progressed about 0.4 of the way to the center.

It is not very likely that the calculation of solidification rates will be of any great practical value in connection with the solidification of metal in molds, where the whole of the metal is allowed to solidify, and where the assumption is made, implicitly in many analyses, that the end of freezing occurs when the solidus temperature is reached. As shown in Chapter 5, almost any alloy will actually solidify over a temperature range that is much larger than the interval between solidus and liquidus. There may, however, be important applications of these methods to the study of problems in which the whole of the liquid is not allowed to solidify, and in which the thickness of the solid layer as a function of time is of real significance.

As an alternative to the analytical methods used by Ruddle, Horvay and Henzel, Adams and others, it is possible, as shown by Paschkis and Hlinka (15), to use an analog computer method for the calculation of the rate of freezing in molds; however, Ruddle (11) suggests grounds for strong criticism of some of the results obtained by Paschkis. To conclude this brief survey of the implications of heat transfer in molds, a few typical results will be introduced in order to give some perspective on the quantities involved. Figure 7.3 shows

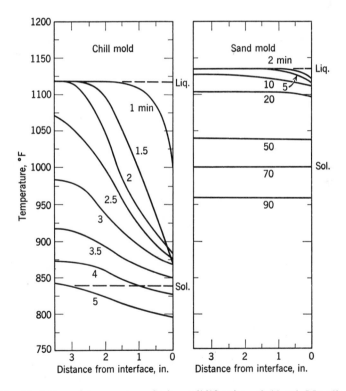

Fig. 7.3. Variation of temperature during solidification of Al 5% Mg alloy in a 7-inch square mold. (a) Metal mold, (b) sand mold. (From Ref. 11, p. 45.)

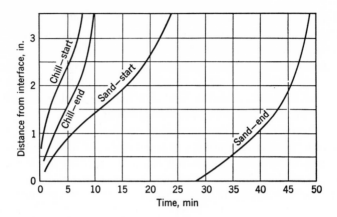

Fig. 7.4. Movement of liquidus and solidus temperatures during solidification of a 0.6% carbon steel. (From Ref. 11, p. 42.)

the temperature-time distribution for a 7-in. sq. section of aluminum 5 per cent magnesium alloy; in a chill (metal) mold, and in a sand mold. Similarly, curves for the location of the liquidus and solidus temperatures, as a function of time, for a 0.6 per cent carbon steel are shown in Fig. 7.4 (17).

Continuous casting. It is characteristic of the traditional casting processes that they are discontinuous, in the sense that essentially the whole of the melt is poured into the mold and allowed to solidify there. The speed of solidification is limited by the heat conduction characteristics of the metal and mold, and a slow speed of solidification is necessarily associated with a large mass of metal. It may be desirable to cast large masses of metal at high speeds, either because of the economic advantages of fast production, or because the properties of the resulting metal are better; the relationship between speed of freezing and structure are discussed in Chapter 8. For these reasons, a variety of processes have been developed in which molten metal is poured continuously into a mold, from which the solidified metal is continuously withdrawn. The rate at which the process can be conducted is determined largely by the problem of the removal of the latent heat and the flow of the metal during solidification, and it is, therefore, appropriate to discuss it here. There are two general types of continuous casting processes; those in which the cross section of the cast metal is completely defined by the mold into which it is cast, and those in which the shape of the cast metal is controlled, at least in part, by the flow of heat into a heat sink. The former type, which has already become important in both steel and nonferrous production, is best fitted for the manufacture of ingots of square, rectangular, or round sections.

The continuous casting processes to be discussed here are shown schematically in Fig. 7.5, in which a mold M, usually of square or circular section, is used to produce the billet B, which is moved downward continuously by mechanical means; usually a reciprocating motion must be superimposed to prevent sticking. Molten metal L is poured into the mold continuously so as to maintain a constant level of liquid in the mold. The mold is cooled, usually by water sprays or jets, and the metal that has emerged from the mold is often also subjected to cooling. The most important characteristic of this process, and the one which distinguishes it most clearly from the discontinuous processes, is that a steady state is maintained. This corresponds to a constant shape of the interface, and a solidification rate that depends only on the position relative to the surface and not on time. The solidification rate is clearly not uniform, because, as shown in Fig. 7.6, solidification at A and B (and in fact at all points)

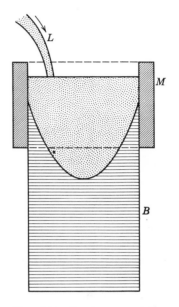

Fig. 7.5. Continuous casting (schematic).

must take place at rates such that the longitudinal components of the rates, AA^1 and BB^1, are equal. The rate of solidification, however, is the rate along the local normal to the interface (BC), which is equal to $BB^1 \sin \phi$, where ϕ is the local inclination of the interface to the direction of motion. It follows that, in order to achieve an interface shape of the kind shown in Fig. 7.6, it is necessary to provide cooling conditions such that the rate of extraction of the latent heat is greatest at A (i.e., on the axis of the billet) and least at D. This requires effective cooling of the part of the billet that has emerged from the mold. If the rate of solidification is a *minimum* at the center of the billet, then the interface will assume a shape of the type shown in

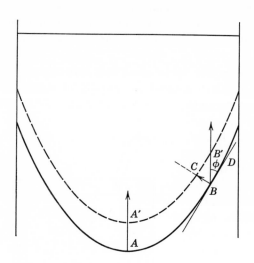

Fig. 7.6. Interface shape and rate of solidification in continuous casting.

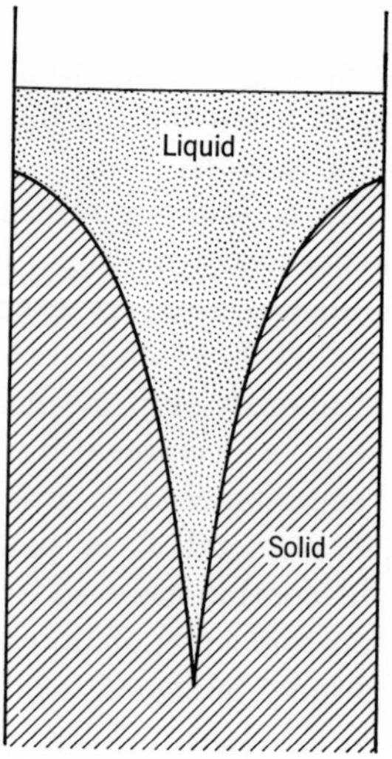

Fig. 7.7. Alternative in-
terface shape in contin-
uous casting.

Fig. 7.7. This is most undesirable because of the excessive accumulation of solute that would occur in the last part to solidify (terminal
transient); strong center-line segregation could not be avoided, and
the possibility also arises that the shrinkage of "sealed off" regions of
liquid near the center line would cause porosity. It follows that
secondary cooling (i.e., below the mold) is necessary.

The quantitative aspects of continuous casting are discussed by
Boichenko (18), who gives the following analysis:

The depth of the liquid core, h_c, is given by the equation

$$h_c = \frac{I\gamma R^2 v_c}{4K(\theta_f - \theta_s)}$$

where I = latent heat plus heat extracted, for unit mass, during fall
of temperature from θ_f to $(\theta_f + \theta_s)/2$
γ = density
R = radius of billet (assumed cylindrical)
v_c = rate of withdrawal of the billet
K = thermal conductivity of the solid metal
θ_f = melting point
θ_s = surface temperature (assumed uniform)

It follows (1) that the depth of core for billets of equal diameter is proportional to the rate of casting (v_c); (2) that the depth of liquid core is proportional to the square of the diameter of the billet; (3) for billets of a given diameter and a given alloy composition, there is a limiting rate of solidification v_l that it is impossible to exceed in practice; this is given by

$$v_l = \frac{4K(\theta_f - \theta_s)}{I\gamma R}$$

The existence of this limit follows from the consideration that as the rate of casting is increased, the length of the liquid core also increases, changing the angle ϕ so that a much greater normal freezing rate is required to give the required longitudinal rates. The limiting rate is imposed in practice by the fact that a very long liquid core produces segregation and porosity difficulties of the kind mentioned in connection with interfaces of reversed curvature. Boichenko quotes experimental data that confirm the validity of the calculation of the liquid core depth; an example is shown in Fig. 7.8, which comes from the work of Dobatkin (19).

It is instructive to compare the characteristics of different metals from the point of view of the depth of the liquid pool that is to be expected. Boichenko uses the expression

$$f(h) = \frac{K(\theta_f - \theta_s)}{L + \frac{1}{2}\gamma_c(\theta_f - \theta_s)}$$

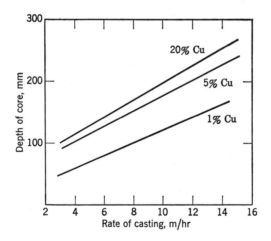

Fig. 7.8. Relationship between depth, liquid core, and rate of casting. (From Ref. 18, p. 146.)

as a parameter that is inversely proportional to the pool depth for a given combination of casting speed and billet size, and he gives the numerical data in Table 7.3.

Table 7.3

Metal	K	L	K/γ_c	$f(h)$
Aluminum	0.53	93	0.82	0.72
Copper	0.92	50	0.98	1.02
Brass	0.28	37	0.33	0.36
Mild steel	0.11	78	0.09	0.10

It can be concluded that the depth of the liquid pool is much greater for steel than for aluminum alloys, for instance, under comparable conditions. A very long liquid pool increases the danger of the liquid breaking out through the solid skin, and the limiting speed or cross section for continuous casting of steel is much less than for other metals. It should also be pointed out that the extreme depth of the liquid pool for steel creates an extra problem: Figure 7.9 shows actual pool depths for steel billets, and it follows that the billet cannot be cut until it has reached a point well beyond the end of the liquid pool, which requires a very tall installation for high-speed casting. It is also of interest to note that, as indicated by the theory, a billet of larger cross section must be cast more slowly, but that the output,

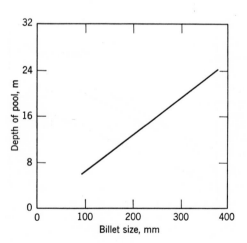

Fig. 7.9. Pool depth for continuously cast steel billets as a function of billet size. (From Ref. 18, p. 152.)

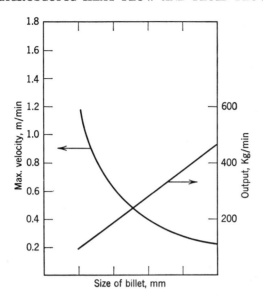

Fig. 7.10. Maximum casting speed and output per unit time as a function of billet size. (From Ref. 18, p. 210.)

in tons per hour, increases with increasing cross section. This is because the volume produced is proportional to the square of the transverse linear dimension, while the speed need be decreased only in proportion to the first power of the size. Figure 7.10 shows how the maximum rate decreases but the output increases with increasing billet size. It will be noticed that it has been implicitly assumed in the preceding considerations of continuous casting that solidification takes place by the formation of a skin and the advance of a smooth solid-liquid interface. This mode of solidification is, as has been pointed out, by no means universal; a serious limitation to the application of the process arises if there is not a continuous skin of solid metal at each point on the casting by the time it has moved downward to a position where it is no longer supported by the mold. The skin must be strong enough to withstand the hydrostatic pressure due to the "head" of liquid metal inside the casting. This problem is accentuated by increasing the speed of the process, and can be overcome by slowing it down.

Dip forming. The second type of continuous casting process, in which the amount and shape of solidified metal is controlled by a heat sink rather than by a mold, forms the basis for the "Dip Forming"

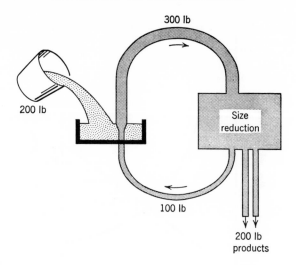

Fig. 7.11.   Principle of dip forming.   (From Ref. 20, p. 12.)

process recently described by Suits (20) and in more detail by Carreker (21).   In this process, a rod of copper, in the form in which the process was first developed, is passed through a bath of molten copper. The copper rod, cold when it enters, is heated up by the bath; the heat required for this comes from the surrounding liquid metal, some of which solidifies on the rod and thereby increases its thickness.   In the example described by Suits, and represented schematically in Fig. 7.11, the cross section of the rod increases, during its passage through the bath, to three times its original value; the rod so formed is then drawn down to the original size and one-third of it is used for further dip forming, while the remaining two-thirds is ready for further processing. The main advantage of this process is that it takes the metal directly from melt to rod, at high speed, eliminating the equipment and time required for pouring ingots and rolling them into rod.   The thermal factors involved in the dip forming process have been analyzed by Horvay (22) for slabs, and by Giaever and Horvay (23) for rods of circular section.   In these papers, the thickness of added metal, assumed to be of the same material as the rod, is calculated as a function of time in the melt, the speed of the bar, and the temperatures, thermal properties, and geometry of the system.   It is shown that the amount of added metal increases to a maximum and then decreases if the depth of the bath is sufficient.   Figure 7.12 shows some experimental points (obtained by Carreker) and theoretical curves using different values of the effective temperatures of the bath and the rod. The experimental points were obtained for speeds ranging from 15 ft/min to 90 ft/min, and for a bath temperature of about 50°F superheat. These results, and many others, show that the process is well understood, and that the analysis provides a good basis for predicting actual performance. However, it is again necessary to point out that the calculations are made on the assumption of a smooth interface. This condition would be satisfied in the case of pure copper, but when the process is applied to an alloy, a dendritic or cellular-dendritic interface is to be expected. It is probable that the maximum amount solidified would be somewhat greater for an alloy, because some liquid would be carried out from the bath in the intra-dendritic spaces; this would solidify subsequently and form part of the solid.   There is no report of full-scale operation of the process on any alloy, although Suits (20) reports that both the General Electric Company and the British Iron and Steel Research Association have made successful experiments on the dip forming of steel.

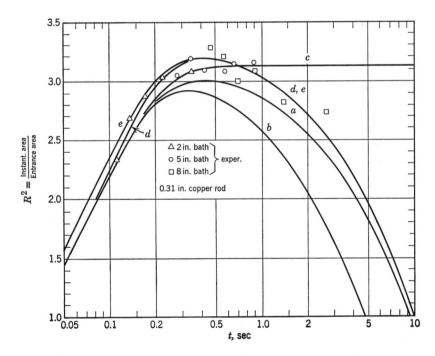

Fig. 7.12. Comparison of experimental and theoretical results of dip forming of copper. (From Ref. 22.)

Horvay and Giaever (24) point out that this analysis of the dip forming process could be applied, with minor modifications, to the casting of one metal on a "heat sink" of another metal. In this case, if the "cast on" metal is to be subsequently removed and used separately, its shape is determined partly by the shape of the surface of the "heat sink" and partly by the amount of metal that solidifies. This type of casting is called, following a suggestion by Fullman (25), a *partial casting*.

7.4 Thermal Stresses in a Solidifying Body

One of the consequences of the macroscopic heat flow that necessarily occurs during solidification is the existence of temperature gradients in the metal that has solidified. These temperature gradients, combined with the contraction of the solid metal as its temperature falls, may give rise to stresses within the solid metal. The general problem of thermal stresses has been considered by Boley and Weiner (26) who have also (27) analyzed the special, and relevant, case of the thermal stresses arising during the solidification of metal in a mold of square section, the metal being assumed to be perfectly elastic up to a yield stress that is zero at the melting point, and increases linearly with decreasing temperature. The metal is, further, assumed to be "perfectly plastic," that is, to deform at its yield stress without work hardening. The thermal conditions in the mold are assumed to conform to the Neumann temperature distribution, as given by Carslaw and Jaeger (28). Weiner and Boley find that the stress history of each "particle" of the solidifying metal is as follows: immediately upon solidification, the metal deforms in tension, since it would be subjected to a tension stress in the plane of the solidifying front if it had a finite yield stress. As it cools, its yield stress increases, but so does the tensile stress on it, so that plastic yielding in tension continues; this stage is terminated when the yield stress exceeds the maximum available stress, and elastic behavior begins. The tensile stress on the "particle" then decreases, becomes zero, and changes to a compressive stress, which eventually reaches a high enough value to cause compressive yielding, which continues thereafter until the analysis becomes invalid. For details of the analysis leading to this rather surprising result, and for an indication of the numerical values that are to be expected, the reader is referred to Weiner and Boley's paper. The thermal stresses discussed above are due to nonlinear temperature gradients in the solid that exist because of the transient nature of the process, and are entirely unrelated to

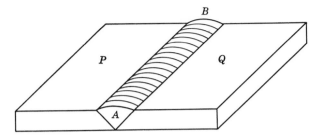

Fig. 7.13. Origin of stresses in a weld.

the thermal characteristics of the mold, except insofar as this determines the rate of heat extraction; there are, however, important cases in which stresses arise as a result of differing thermal effects in the solidifying metal and in the material with which it is in contact. An example of the generation of high stresses in this way is in a welded joint between two metal plates, shown schematically in Fig. 7.13, in which molten metal is introduced into the groove AB between the two plates P and Q. Unless the plates are pre-heated, the "bead" contracts far more, after it is solid, than do the plates, and high stresses are generated in and near the weld. Plastic deformation may relieve the stresses, but cracking often occurs instead.

A second example of stresses that arise as a result of contraction of the solidified metal is to be found in castings which are shaped so that the mold restrains the contraction of the metal. A simple example is shown in Fig. 7.14, in which the contraction of the mold between A and B may be insufficient to relieve stresses; if the alloy solidifies over a long temperature range, the contraction of the first part to solidify may tear the metal apart where there is still some liquid, resulting in extreme weakness, although the casting may appear superficially to be sound. This is an example of the phenomenon known as "hot-tearing," which apparently occurs while there is still some liquid present.

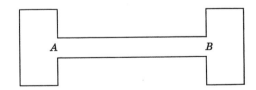

Fig. 7.14. Origin of contraction stresses in a casting.

References

1. E. N. da C. Andrade, *Proc. Phys. Soc.*, **52**, 748 (1940).
2. E. N. da C. Andrade, *Proc. Roy. Soc. A*, **215**, 36 (1952).
3. E. N. da C. Andrade, *Proc. Phys. Soc. B*, **63**, 773 (1950).
4. T. P. Yao and V. J. Kondic, *J. Inst. Met.*, **81**, 17 (1952).
5. W. R. D. Jones and W. I. Bartlett, *J. Inst. Met.*, **81**, 145 (1952).
6. G. M. Panchenkov, *Dokl. Akad. Nauk SSSR*, **79**, 985 (1951).
7. B. R. T. Frost, *Progr. in Metal Phys.*, **5**, 102 (1954).
8. *Metals Handbook*, American Society for Metals, Cleveland, 1948, p. 6.
9. *Metals Handbook*, American Society for Metals, Cleveland, 1948, p. 201.
10. G. Horvay and J. G. Henzel, *Trans. Met. Soc. AIME*, **215**, 258 (1957).
11. R. W. Ruddle, *The Solidification of Castings*, Institute of Metals, London, 1957.
12. R. W. Ruddle and A. L. Mincher, *J. Inst. Met.*, **78**, 229 (1950–1951).
13. W. S. Pellini, *Trans. AIME*, **210**, 136 (1957).
14. C. M. Adams, Jr., *Liquid Metals and Solidification*, American Society for Metals, Cleveland, 1958, p. 187.
15. V. Paschkis and J. W. Hlinka, *Trans. Am. Foundrymen's Soc.*, **65**, 222 (1951).
16. F. A. Brandt, H. F. Bishop, and W. S. Pellini, *Trans. Am. Foundrymen's Soc.*, **62**, 646 (1954).
17. H. F. Bishop, F. A. Brandt, and W. S. Pellini, *Trans. Am. Foundrymen's Soc.*, **59**, 435 (1951).
18. M. C. Boichenko, *Continuous Casting of Steel*, Butterworth's, London, 1961.
19. V. I. Dobatkin, *Metallurgy*, **14**, 101 (1948).
20. C. G. Suits, Charles M. Schwab Memorial Lecture, American Iron and Steel Institute, May 1963.
21. R. P. Carreker, *J. Met.*, **15**, 774 (1963).
22. G. Horvay, Unpublished report.
23. I. Giaever and G. Horvay, Unpublished report.
24. G. Horvay and I. Giaever, Unpublished report.
25. R. L. Fullman, quoted in Ref. 21.
26. B. A. Boley and J. H. Weiner, *Theory of Thermal Stresses*, John Wiley & Sons, New York, 1960.
27. J. H. Weiner and B. A. Boley, *J. Mech. Phys. Solids*, **11**, 145 (1963).
28. H. S. Carslaw and J. C. Jaeger, *Conduction of Heat in Solids*, 2nd Ed., Clarendon Press, Oxford, 1959.

8

The Structure
of Cast Metals

8.1 General Considerations

In the preceding chapters an attempt has been made to examine all the different physical processes that take place during solidification; in each case, the experiments that are most relevant are those which most completely isolate one particular aspect of the problem, and the theoretical discussion and conclusions are, either explicitly or implicitly, based on the premise that a single process is under consideration.

The purpose of this final chapter is to examine the application of our understanding of the various distinct processes to practical situations in which the objective is to produce "hardware" rather than understanding. There are two phases of the problem of applying our knowledge to practical problems; it is first necessary to understand and explain the observations that have been made, either in actual practical cases or in the "simulated practice" type of experiment; that is, the type of experiment in which the complexity of the system is no less than in practice, but the variables are under better control. The second phase is to use our understanding to predict what will happen in hitherto untested conditions and hence to point the way to processes that are better than existing ones in terms of either quality or economy. It is obvious that the latter aspect of applications must always lie in the future, and that it depends on skill, experience, and intuition as well as on knowledge; the objective is to show how our understanding of the separate aspects can be applied to complicated situations.

The problem, therefore, is to explain the origin of all the different features that together constitute the "structure" of cast metal, and to show how these structural features depend on the physical and chemical variables of the process. This should suggest where to look if the structure is not what is wanted, and will indicate which variables are critical and which are not. The "structure" of the cast metal, when the term is used in its broadest sense, includes the "structure" in the metallographic sense (i.e., the size, shape, orientation, and perfection of the crystals), the distribution in the metal of the various chemical elements that are present (i.e., segregation, both "macro-" and "micro-") and the internal and external topography of the metal (porosity, soundness, and surface shape and finish).

The main variables that account for the differences between the structures of different samples of cast metals are the chemical composition of the alloy and the rate at which solidification took place; but to be realistic, it is necessary to include in "composition" the gases which may evolve during solidification and the impurities which may be responsible for nucleation, as well as the elements that are included in the nominal composition of the alloy; and the rate of solidification must be broadened to take account of the fact that different parts of the metal freeze at different rates and that the liquid metal may not be stationary during solidification.

It is convenient to subdivide all the useful types of solidification process into three main categories, as follows: (1) those in which the product is to be used in essentially its cast form; these are described as "castings," (2) those in which the metal is cast into such a form that it is suitable for further processing by plastic deformation; these are ingots and billets, and (3) those in which the cast metal forms part of a more complicated structure; this category includes welds, brazed and soldered joints, and hot dipped and sprayed coatings.

In the first category, that of castings, the structure that is produced by the solidification process is preserved, except insofar as it can be changed by heat treatment; in the case of steel, very profound structural changes can be produced by transforming the steel into austenite and then transforming it back by controlled cooling, which may be followed by further heating; in most other alloys, a completely new structure cannot be obtained, although age-hardening can be used to improve the properties without causing any change in the size, shape, or orientation of the crystals. The other change that can be produced in a casting is homogenization, that is, the "smoothing out" of short-range concentration gradients by diffusion. It follows that, except perhaps for steel castings, the structure that is formed on solidification controls the properties of the final product, and even in the case of steel, the final product "inherits" some of its structure and properties from the structure that forms on solidification.

Thus porosity, segregation, undesirable grain size or preferred orientation must either be controlled during solidification or accepted in the final structure. In the ingot and billet category, the structure formed during solidification is changed by subsequent deformation and annealing; this produces an entirely new grain structure but it does not completely "cure" either porosity or segregation, both of which may have serious effects on the properties of the material even after drastic deformation such as rolling into sheet form.

The decision to adopt a particular process is often a compromise between economics and technical desirability; for example, for a short run, it might be much cheaper to use wooden patterns and a sand mold than a permanent (metal) mold. One result of using a sand mold is that the rate of extraction of heat is much lower, because of the low thermal conductivity of sand compared with that of iron. It would be useless to attempt to describe the effects of all possible changes in all the process variables, including the composition of the alloy as well as the mold materials, pouring temperature, mold temperature, and geometry, and we will instead discuss the sequence of events that take place and how the result depends upon the appropriate variables. For this purpose, it is convenient to discuss, in order, the aspects of structure that relate to crystal size and shape, to segregation, and to the volume changes that accompany solidification.

8.2   The Macrostructure of Cast Metals

Description of cast structures.   When a metal or alloy is poured into a mold and allowed to solidify, it may do so in various ways, depending mainly on the rate of heat extraction, the amount of metal and its composition and the potency of the nucleants that are present. At one extreme, a relatively small mass of a metal or an alloy, cooled rapidly, will solidify as a large number of equiaxed crystals; that is, crystals that are not elongated in shape. At the other extreme, a large mass, cooled slowly, will consist at least partly of columnar crystals; that is, crystals whose length, perpendicuar to the mold wall, is many times their other dimensions. Figure 8.1 shows examples of (1) a completely equiaxed structure, (2) a completely columnar structure, and (3) a partially columnar structure. The completely equiaxed structure is usually regarded as being characteristic of "castings," while a partially columnar structure is referred to as an "ingot structure." Metallographic studies have shown that

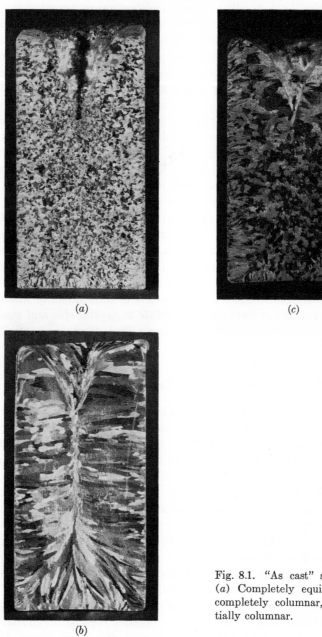

Fig. 8.1. "As cast" structures. (a) Completely equiaxed, (b) completely columnar, (c) partially columnar.

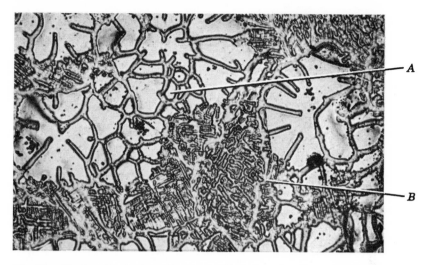

Fig. 8.2. Equiaxed grain in an ingot of Al-2% Cm, showing pre-dendritic region (*A*) and cellular dendritic region. (Photograph by H. Biloni.)

both columnar and equiaxed crystals start growing pre-dendritically and that this develops into cellular dendritic growth. Figure 8.2 shows an example of an equiaxed grain, in which the region *A* has the characteristic appearance of pre-dendritic growth, while the region *B* is cellular dendritic.

The two main characteristics by which a structure can be described are the extent, if any, of the columnar zone, and the grain size of the equiaxed region. These two characteristics, which can be expressed quantitatively, depend upon the process variables in a way not yet analyzed quantitatively except in a few special cases; this is because so many different aspects of the process can vary. The process variables can be divided into two groups, namely, those that are related to the metal, and those that describe the mold. The metal is characterized by its composition, its nucleation characteristics, its temperature when it is poured, and its motion as it enters the mold. The mold is characterized by its thermal properties, temperature and geometry.

Experimental observations. Although a great many observations have been made on the effects of process variables on the two structural parameters, most of them are of little use from the point of view of understanding the physical processes that control the structure. Some recent experiments (1), however, throw new light upon this old problem. In these experiments, aluminum copper alloys were cast in a standard graphite mold at a series of pouring temperatures, and the columnar

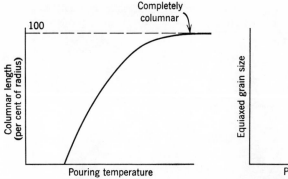

Fig. 8.3. Variation of length of columnar zone with pouring temperature.

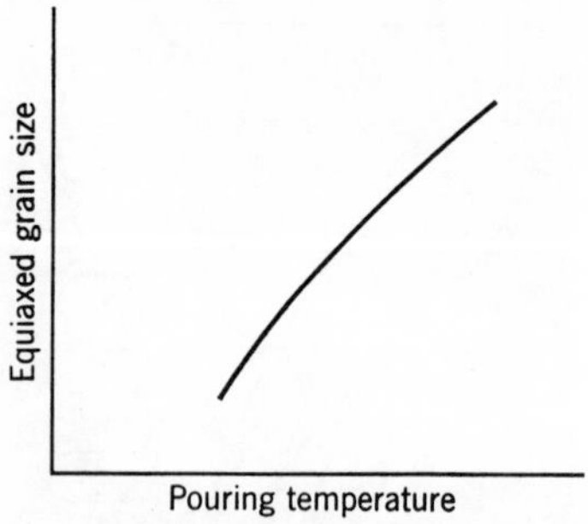

Fig. 8.4. Variation of equiaxed grain size with pouring temperature.

length and equiaxed grain size were measured. The results provide a description of the effects of some of the process variables:

(a) For a given alloy and mold, the columnar zone increases in length as the pouring temperature is raised (Fig. 8.3).

(b) Under the same conditions, the equiaxed grain size decreases as the pouring temperature is lowered (Fig. 8.4).

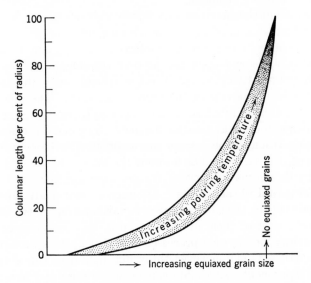

Fig. 8.5. Variation of equiaxed grain size with columnar length.

(c) For the same conditions as (a), the grain size increases as the columnar length increases (Fig. 8.5).

(d) For a pouring temperature that is a fixed amount above the liquidus, the columnar zone decreases in length as the alloy content is increased (Fig. 8.6).

These experimental results were obtained with rather small scale experiments (graphite mold, height 3 in., diameter 2 in.) but they are in qualitative agreement with many other observations. They can be used as the basis for an explanation of the sequence of processes that occur during solidification in a mold.

Solidification in a mold. CHILL ZONE. When the metal enters the mold, which is far below the melting point of the metal, a layer of metal that is in contact with the mold wall is chilled to such a degree that copious nucleation takes place in it. The number of nuclei that form during the initial chill depends also upon the effectiveness of the nucleant particles present. If they nucleate at very

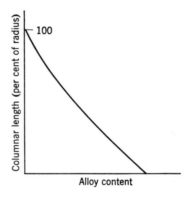

Fig. 8.6. Variation of columnar length with alloy content; constant superheat.

small supercooling, then each crystal grows slowly, gives out little latent heat, and so allows other nucleants in its immediate vicinity to operate even if they are slightly less effective than the first one; on the other hand, if considerable supercooling is required for nucleation, then each crystal that forms grows very fast and suppresses all other nucleants in a considerable volume around it. Thus for a given rate of heat extraction, the number of nuclei that form increases as the supercooling required for nucleation decreases. Further, the number that nucleate increases as the rate of heat extraction increases. These considerations assume that the number of nuclei that form is not limited by the number of available nucleant particles, and this is almost always true. The number of crystals that would be nucleated in the chill zone, therefore, depends upon the effectiveness of the nucleants, the rate of heat extraction, and the volume of chilled liquid. The two latter quantities depend upon the temperature of the mold surface when the metal reaches its liquidus temperature and on the thermal properties of the mold and the metal; the higher the pouring temperature, the more the mold will have been heated while the metal at the surface is being chilled, and the slower will be the rate of heat extraction during the critical time when the nucleation temperature is reached; a low pouring temperature, therefore, allows many crystals to nucleate, while a high pouring temperature reduces the number. Similarly, a mold which has a high thermal diffusivity removes heat fast and causes a greater nucleation density in a greater volume of liquid. Low thermal diffusivity of the liquid metal has a similar effect. The effect of composition must also be considered; the rejection of solute decreases the speed of dendritic growth at a given amount of supercooling, and therefore increasing alloy content should allow more nuclei to form.

Some of the initially formed crystals may nucleate on the mold wall and be in good thermal contact with it; some, in general, will not. If the mold wall is less effective as a nucleant than the particles in the liquid, then nucleation will occur only in the liquid and not on the wall. The nuclei that form in the initially chilled zone grow, at first, pre-dendritically, and the temperature of the remaining liquid very quickly falls to the temperature of the solid-liquid interface of the growing crystals. This temperature depends upon the rate of growth, which in turn is controlled by the rate of conduction of heat to the mold. If this temperature is above the nucleation temperature of the melt, no further nuclei are formed, and an "ingot" type of structure is produced. If, on the other hand, the growth temperature for the imposed rate of heat extraction is below that required for nucleation, then nucleation takes place throughout the melt and a "casting" type of structure is formed.

COLUMNAR ZONE. The columnar structure consists of crystals that nucleated on or near the mold wall and grew inward as a result of heat flow into the mold. The interface between the columnar crystals and the liquid is smooth, cellular, or cellular-dendritic according to the composition of the alloy and the rate of solidification; for all except rather pure metals it is cellular-dendritic. The crystals that were nucleated on the mold wall, or sufficiently close to it to be "trapped" in the early stages of solidification, grow in competition with each other and a preferred orientation develops as a result of the suppression of those crystals that are least favorably oriented for growth away from the wall; the crystals that are formed initially are usually randomly oriented, but in some cases they may have a preferred orientation as a result of epitaxial nucleation on the mold.

During the growth of the columnar zone, the number of crystals decreases, the cross section of those remaining increases, and a preferred orientation is developed by the selection for survival of crystals which are favorably oriented. The favorable orientation is always that in which the direction of dendritic growth is perpendicular to the mold wall ($\langle 100 \rangle$ for face-centered cubic and body-centered cubic, $\langle 10\bar{1}0 \rangle$ for hexagonal close packed). The mechanism of selection and the suppression of the unfavorable orientation has been studied by Walton and Chalmers (2) who concluded that the initial period of dendritic growth favored the crystals that happened to nucleate with the right orientation. This can occur in 2 ways. In the first place, dendritic growth along the mold wall will be more rapid if there are dendrite directions parallel to the wall. This occurs fully, in the cubic structures (in which there are 3 mutually perpendicular dendrite directions), only when there is also a dendrite direction normal to the wall. The favored crystals therefore grow laterally on the wall faster than their com-

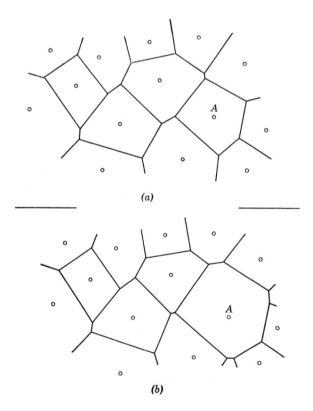

(a)

(b)

Fig. 8.7. Shapes of crystals growing from randomly selected points. (a) Equal rates, (b) crystal A growing twice as fast as the others.

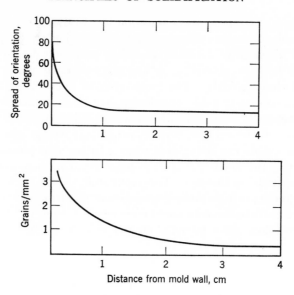

Fig. 8.8. Variation of grain size and preferred orientation with distance from mold wall. (From B. Chalmers, *Physical Metallurgy*, John Wiley and Sons, New York, 1959.)

petitors. They also grow away from the wall, in the initial period of dendritic growth, faster than other crystals, allowing their transverse branches to spread laterally ahead of those of the less favorably oriented crystals.

As a result of either or both of these processes, the favorably oriented crystals are larger than their competitors when columnar growth begins. Their larger size, together with their greater average number of sides, gives them an advantage during the columnar stage of growth. This can be explained as follows: Fig. 8.7a represents a cross section, parallel to the plane of the mold wall, of an array of crystals that originated from randomly spaced points, and grew outward at the same speed in all directions (each boundary is the perpendicular bisector of the line joining adjacent nucleation sites). Figure 8.7b represents the same nucleating points, but with the crystal A having grown twice as fast as the others. It will be seen that the crystal that has grown faster is not only larger, but that it also has more sides, in this case 8, compared with the average number of 6. The crystals with more sides tend to increase in size because the equilibrium dihedral angle (120°) between faces can be maintained by tilting the 3-grain intersection outward as the crystal grows upward (away from the mold wall). The increase in size of the larger crystals, with the consequent elimination of the smaller ones, progressively decreases the total extent of the grain boundary in successive sections of the growing crystals. This argument is essentially similar to that of C. S. Smith (3) for the growth of many-sided crystals at the expense of those with fewer sides in the process of grain growth in solids. The elimination of the unfavorably oriented grains during the columnar stage of growth is gradual; the increase of average grain size and the sharpening of the preferred orientation develop together, as shown in Fig. 8.8.

If because of a high pouring temperature, the initially formed "chill" crystals were remelted, new crystals are formed when the part of the melt in contact with the wall again reaches the nucleation temperature. The cooling is much slower now than when it impinged on the cold wall of the mold, and hence far fewer crystals are nucleated. This is because the first few have time to grow outward along the mold wall to a considerable extent before the temperature falls far enough anywhere to nucleate any more crystals. The crystals that are formed, therefore, are few in number, and probably of random orientation. Their subsequent growth is far less competitive than that in the case discussed before, and the result is that the crystals that grow inward from the mold wall do not exhibit any preferred orientation.

The columnar zone, therefore, may consist of the crystals from the chill zone (consisting of the small crystals that nucleated on the wall) that have survived the competitive growth process on account of their favorable orientation; the chill zone may, however, be missing, as a result of sufficient superheat to prevent initial nucleation or to remelt

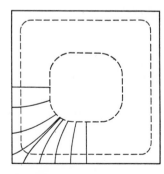

Fig. 8.9. Grain boundaries orthogonal to isotherms (schematic).

Fig. 8.10. Grain boundaries maintaining original directions (schematic).

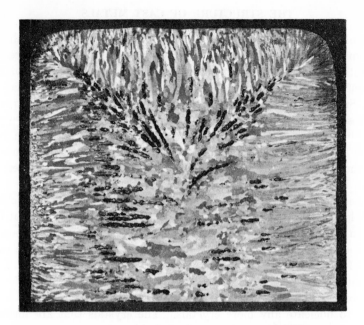

Fig. 8.11. Photograph of structure shown in Fig. 8.10.

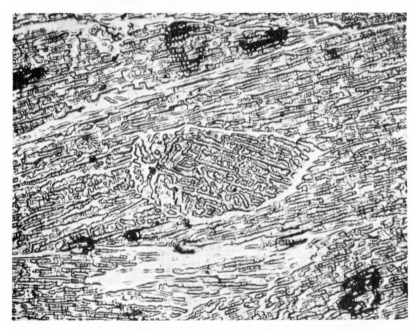

Fig. 8.12. Equiaxed crystals in columnar zone. (Photograph by H. Biloni.)

the chilled crystals. In this case the columnar zone consists of fewer, larger crystals with no preferred orientation. In either case, there is ample evidence that the columnar zone is cellular-dendritic in alloys that have an appreciable freezing range.

It has been shown (4) that the grain boundaries of a structure growing with a smooth front are always nearly perpendicular to the solid-liquid interface. In this type of growth, therefore, the boundaries near the corner of a mold should appear as shown in Fig. 8.9, where 2 positions of the solid-liquid interface are shown. In many actual cases, however, the structure is actually that shown in Fig. 8.10; an example is shown in Fig. 8.11. This structure is a result of the strong crystallographic character of the cellular-dendritic type of growth. It is generally believed that the columnar zone consists entirely of crystals that nucleated at the mold wall; however, it has recently been shown (5) that (a) a considerable number of small equiaxed crystals can be found in the columnar zone. Figure 8.12 shows a typical region in an aluminum 4 per cent copper alloy. It is significant that the small equiaxed crystals have a pre-dendritic structure and (b) that some of the elongated crystals originate, in the columnar zone at points well removed from the vicinity of the cold wall (see Fig. 8.13).

EQUIAXED ZONE. Two theories have been advanced for the origin of the equiaxed zone; the first, due to Winegard and Chalmers (6), proposed that nucleation of the crystals of the equiaxed structure occurs when the liquid reaches its nucleation temperature, as a result of constitutional supercooling. The subsequent rapid growth of the new crystals would prevent further growth of the columnar grains. The alternative proposal (1) is that nucleation occurs only during the initial chilling, the equiaxed zone consisting of crystals that survived and grew in the liquid until they formed a network that inhibited further growth of the columnar zone. The evidence in support of this theory is as follows: (a) The correlation between columnar length equiaxed grain size and pouring temperature is accounted for; a high pouring temperature gives a "thinner" zone in which nucleation can take place, hence fewer "free" chill crystals, combined with a higher probability of the "free" crystals being remelted before the temperature of the liquid falls to a level at which they can grow. There are, therefore, relatively few "free" crystals; they survive and grow for a relatively long time before impinging on each other and forming a continuous network. Conversely, a low pouring temperature produces many nuclei of which a high proportion survive, giving a continuous network early in the process, with the result that the columnar zone is short and the equiaxed grain size is small. (b) There is no evidence

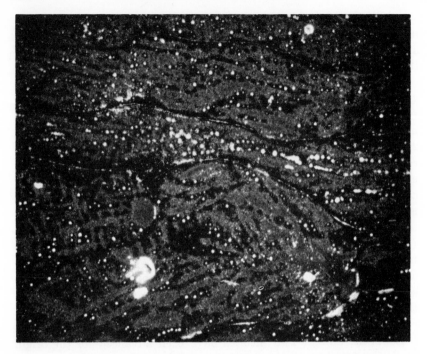

Fig. 8.13. Columnar crystal originating in columnar zone. (Photograph by H. Biloni.)

of a continuous fall in temperature in the center followed by a rise after the columnar zone has started to grow as would be expected on the earlier theory. (c) The presence in the columnar zone of small equiaxed crystals is predicted by the newer theory and not by the older one. (d) The following experiment (1) shows that the equiaxed zone is not formed purely as a result of temperature changes in the liquid. An alloy was poured into a mold in which the center region was isolated mechanically, but not thermally, from the outer region; conditions were selected so that a short columnar zone and a fine equiaxed structure were produced in the outer part of the mold. The inner part solidified with an extremely coarse structure (Fig. 8.14). (e) Walker (7) has shown that a nickel, 20 per cent copper alloy containing no nucleant particles solidifies with typical chill, columnar, and equiaxed structures, A, B, and C, when cast under suitable conditions (Fig. 8.15), while pure nickel, cast under the same conditions, gives a purely columnar structure.

It should be pointed out that an equiaxed zone can be produced by constitutional supercooling; for example, a detailed mathematical analysis has been made by Tiller (8) of the development of constitutional supercooling in a semi-infinite liquid into which a solid-liquid interface advances at a rate that is proportional to the square root of time.   This analysis shows that constitutional supercooling would develop, and that rapid nucleation would occur, after columnar growth had reached a length determined by the characteristics of the system. This analysis describes a specific situation which in no way resembles the solidification of a metal in a mold of limited extent, in which the fall in temperature of the liquid is very rapid compared with its rate under the conditions selected by Tiller.   It is concluded that this analysis throws little light on the formation of equiaxed grains in an ingot.

Fig. 8.14.   Structure produced when the inner region is mechanically isolated from the outer region.

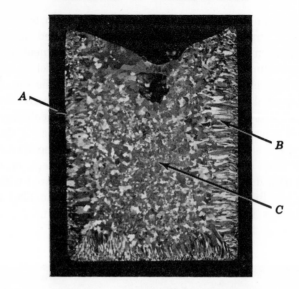

Fig. 8.15.   Ingot structure in nucleant-free nickel copper alloy.   (Photograph by J. L. Walker.)

In addition to the two theories for the origin of the equiaxed zone discussed above, it has also been suggested that the equiaxed crystals are formed by the growth of detached fragments of the columnar crystals; both mechanical and thermal mechanisms of detachment have been proposed, but there is no evidence that there is sufficient fluid motion or heterogeneity of temperature in the liquid.

Thus it is concluded tentatively that the columnar zone is terminated by the already existing skeleton of the equiaxed zone or, more precisely, by encountering liquid already enriched with solute by the growth "free" crystals; the higher the pouring temperature the longer the superheat persists and the fewer initially nucleated grains survive because most of them move into the hotter liquid and melt; those that do survive have time to grow to a substantial size before the columnar zone reaches them; the growth of the columnar zone itself is slower with higher pouring temperatures because the mold has been heated up by the superheat of the liquid metal.

EFFECTS OF GRAVITY AND OF ROTATION.   Another type of mixed structure described by Ruddle (10) is that in which a columnar region is formed at the top while the lower part is equiaxed (Fig. 8.16).   This is attributed to the "settling out" of the equiaxed crystals after they have nucleated, so that the columnar zone at the top is not terminated by the presence of the equiaxed grains. However, it should be recognized that in this example, the solute enriched liquid also has a tendency to sink, thereby reducing the constitutional super-cooling at the upper part of the interface between the columnar zone and the melt. A less extreme form of the same phenomenon is illustrated in Fig. 8.17 in which a completely equiaxed zone shows a gradation of grain size from coarse at the top to fine at the bottom. This is probably caused by the "settling" of growing crystals in the liquid.

If these conclusions are correct, they should account for the effects that have been observed when the liquid is set in rotation within the mold by electromagnetic stirring. The basic observations are: (a) that rotation of the liquid as it is solidifying causes grain refinement and shortens the columnar zone, and (b) that the columnar zone is

Fig. 8.16. "Mixed" structure due to "settling" of crystals.

Fig. 8.17. Variation of equiaxed grain size with position in the ingot.

inclined in the "upstream" direction to the interface. Both of these effects are shown in Figs. 8.18 and 8.19, taken from a paper by Roth and Schippen (11). In Fig. 8.18 the rotation was continuous while Fig. 8.19 shows the effect of reversing the direction of rotation.

Interesting additional observations are:

(a) that of Crosley, Fisher, and Metcalfe (12) that if solidification does not begin until the steady-state rotation speed of the liquid has been reached, the normal structure for a static casting is obtained, but that periodic reversal of the direction of rotation produces a fine-grained structure. The explanation offered by Crosley, Fisher, and Metcalfe is that nucleation is enhanced by viscous shear in the liquid, and they show that under their conditions of periodically reversed stirring, very high shear rates are obtained.

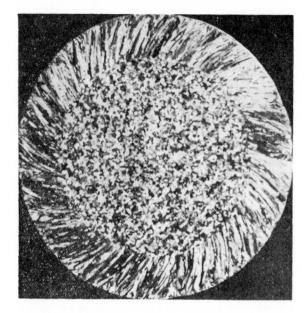

Fig. 8.18. Effect of rotation of the liquid on grain size and columnar length. (From Ref. 11.)

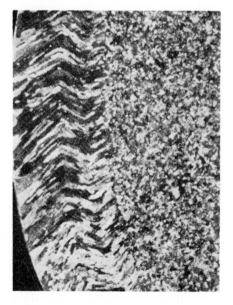

Fig. 8.19. Effect of rotation on the columnar structure. (From Ref. 11.)

(b) It is shown by Langenberg, Pestel, and Honeycutt (13) that if rotation is allowed to stop during solidification of a type 310 stainless steel, the structure reverts to the columnar form.

The major conclusions to be drawn from these observations are that an equiaxed structure can be produced in an alloy at any stage of solidification by introducing shear motion between the liquid and the solid. This may be caused by the removal of dendrite tips by the viscous forces that are set up, or perhaps by the centrifuging action of the rotation on the crystals already growing in the liquid. If they are forced outward against the surface of the columnar zone, the equiaxed zone is formed. The fact that a finer grain is produced in this way than by static cooling may be because of the more rapid cooling of the liquid resulting from its better heat transfer to the columnar zone. The absence of a refining effect when rotation is already established when freezing starts may also be explained on the basis of centrifugal effects which prevent the initial nuclei from ever leaving the vicinity of the mold wall. The long columnar zone that is often found in centrifugal castings (in which the mold is rotated about its own axis) also points to the importance of centrifugal effects; but the possible significance of the fragmentation theory must not be overlooked. The inclination of the columnar boundaries is of interest mainly as a way of determining the direction of fluid flow during solidification by subsequent study of the solid metal. When the flow of the liquid is slow, perhaps laminar, the columnar zone is extended beyond its usual length, and the columnar boundaries are inclined to the interface in such a way as to point "upstream." It appears, therefore, that there are two almost opposite effects than can occur as a result of flow of the liquid. The explanations may be as follows. If the liquid flow across the surface reduces the boundary layer sufficiently, the dendritic structure cannot form; the interface is cellular, and the profile of the cells should be modified by the combination of transverse diffusion with the transverse flow of the liquid. The effective diffusion distance should be increased in the direction of flow (AB, Fig. 8.20), and decreased in the opposite direction (AC). The leading point of the cell A is the point from which the diffusion is equal in both directions, and it will move toward C as the flow increases in speed. The result will be a tilt of the cell walls in the direction shown in Fig. 8.20.

It is probable, but not confirmed by experiments, that the columnar grain boundaries are steered by the cell walls, as has been found to be the case for low-angle lineage boundaries. This would account for the observed deviation of the columnar boundaries.

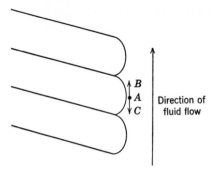

Fig. 8.20. Effect of transverse fluid flow on diffusion.

THERMAL MEASUREMENTS. Another way of looking at the process of solidification in a mold is that developed by Pellini and his co-workers (14), by Ruddle (15), and by many other workers. The basic experimental technique used by these investigators is to measure the temperature as a function of time at a series of points on and within

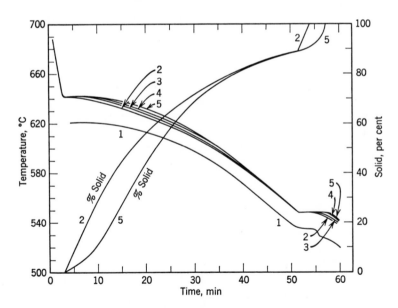

Fig. 8.21. Cooling curves for a sand casting of aluminum, 4% copper, diameter, 5 in.; length, 10 in. Position of thermocouples: (1) interface (sand side), (2) interface (metal side), (3) 1 inch from interface, (4) 2 inches from interface, (5) 5 inches from interface (center). (From Ref. 10, p. 44.)

the solidifying metal. These elegant experiments, examples of the results of which are shown in Figs. 8.21, 8.22, and 8.23, demonstrate clearly that the superheat always disappears quite quickly and that, whereas large temperature gradients can exist in a chill casting while it is solidifying, this is not true for a sand casting. The interpretation of the experimental results is, however, open to some doubt, since it is based on the concept that the liquid remains unchanged in composition throughout, and that the equilibrium diagram can be used to determine the proportion of material that has solidified from its temperature. Following this point of view, both Ruddle and Pellini interpret the process of solidification as the progression through the liquid of a "nucleation or start of freezing wave" followed by an "end of freezing wave." The "start of freezing wave" represents the time at which the liquid reaches a temperature that is described as the liquidus temperature. However, it is clear that in an alloy, both

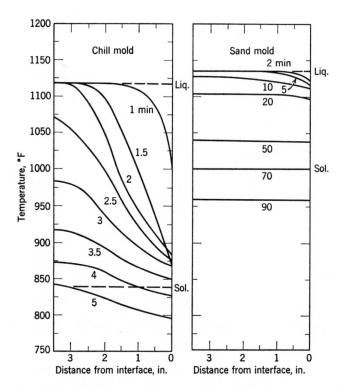

Fig. 8.22. Temperature-distance-time curves for a 7-in. square section in Al–8.5% Mg. (From Ref. 10, p. 45.)

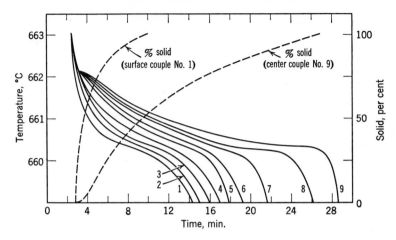

Fig. 8.23. Cooling curves for an ingot of super purity aluminum with 0.11% titanium. (From Ref. 10, p. 47.)

nucleation and dendritic solidification must take place at temperatures that are below the liquidus temperature of the original liquid, which in any case has little relevance once the composition is changed, even locally, by the rejection of solute. The relevance of the solidus temperature (end of freezing wave) is even more dubious, as the process of solidification does not terminate at the solidus temperature except when the terminal liquid is of a eutectic composition (see Chapter 5).

GRAIN REFINERS. As indicated in Chapter 2, the theory of heterogeneous nucleation is not altogether satisfactory; while it is usually possible to explain why a particular substance is effective as a nucleant, one cannot, in general, predict how effective a material will be. It is therefore possible only to report some empirical findings which seem to be supported by adequate evidence.

Aluminum Alloys. It is recognized that the most effective grain refiner for aluminum and many aluminum alloys is titanium, and it is believed that either titanium carbide or titanium nitride forms the actual nucleation catalyst (16, 17). However, it has also been found that niobium, presumably present as the carbide, is also effective as a grain refiner (18).

Magnesium Alloys. Many magnesium alloys can be grain-refined by superheating to about 850°C and then rapidly cooling to the casting temperature and pouring immediately. The most acceptable explanation of the grain-refining effect of superheating is the following, given by Warrington (19). There are three compounds of manganese and aluminum, $MnAl_6$, $MnAl_4$, and $MnAl_3$; $MnAl_4$ is stable between 715 and 850°C and has a hexagonal structure, with nearly the same lattice parameters as magnesium, but the structures of the two other phases ($MnAl_6$ and $MnAl_3$) are quite different. Thus it seems that the effect of superheating to 850° is to form the required nucleant particles. However, it was suggested by Baker, Eborall, and Cibula (20) that the nucleating agent in this case was aluminum carbide. It is also found that magnesium alloys can be grain-refined by means of finely divided carbon (in the form of lampblack) and by zirconium, and by titanium unless manganese silicon compounds are formed (21).

Copper Base Alloys. Dennison and Tull (22) have shown that grain refinement depends on the nature of the phase that nucleates first; this depends on the composition of the alloy. The copper aluminum system has a eutectic at about 8 per cent aluminum; an alloy with 7 per cent Al, which should form the α phase first, can be grain-refined by the addition of 0.1 per cent of Mo, Nb, W or V, while a 9 per cent alloy is refined by 0.02 per cent Bi, but not by the previously mentioned elements.

Tin. It was shown by Pokrovsky and Tissen (23) that 0.4 atomic per cent of germanium gives fine-grained tin at all cooling rates, while the same concentration of indium produced the same effect at fast cooling rates, but not at slow ones.

Type Metals. Arsenic (0.1 per cent) and tellurium (0.15 per cent) drastically reduce the grain size of type metals (24).

Structure of continuous castings. It is generally agreed that it is desirable, for castings and for ingots, to minimize segregation, grain size and porosity; all these characteristics are improved by increasing the rate of solidification, which can be achieved, in castings, only by increasing the conductivity of the mold wall and by decreasing the size of the casting; for ingots the efficiency of cooling can be greatly increased by adopting the geometry characteristic of a continuous casting process.

It should be pointed out that the physically important parameters in the solidification process are the rate of advance of the isotherms and the temperature gradient in the liquid, since these two jointly determine the extent of the constitutional supercooling that would occur if the interface remained planar. The parameter that is most readily measured in practical situations, however, is the rate of fall of temperature at a given point. Although this parameter can be computed easily from the rate of advance and the temperature gradient, the inverse is not true. Consequently there is relatively little information on the effect of the freezing rate on the grain size in castings, and, while it is generally believed that a higher cooling rate produces a finer grain size, this is apparently not universally true. For example, Green (25) found that the grain size of a magnesium base alloy casting decreases as the cooling rate is increased from 200 degrees Fahrenheit per minute up to 600 degrees per minute, beyond which it remains constant; it has also been reported that, in some cases, grain size is independent of cooling rate (26). It is found, however, that when the rate of solidification (as distinct from the rate of cooling) is measured, the grain size does decrease with increasing rate. The most suitable system for making this kind of comparison is a continuous casting, in which a steady-state rate of advance of the interface is achieved. Dobatkus (27), for example, gives the relationship shown in Fig. 8.24 between rate of solidification and "dendritic grain size" for duralumin. It should be pointed out that, although the rate of solidification is a significant parameter, the temperature gradient in the liquid also varies; it must increase with increasing speed of solidification under these conditions.

An important limitation to the speed with which continuous casting can be carried out is the necessity to establish a skin of solid metal that is strong enough to contain the liquid by the time the support given by the mold is no longer available, that is, by the time the metal moves downwards out of the mold. It is not difficult to calculate the

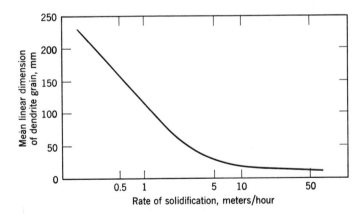

Fig. 8.24. Relation between rate of solidification and "dendritic grain size" for duralumin. (From Ref. 24.)

profile that the surface of the solid would have if the metal were pure; this is the case in which there is a sharp interface (no dendrites) and all the solid is outside this surface. However, this calculation is of little value if the interface may be cellular-dendritic, and the question that arises is whether the boundary of the completely solid region can be found either by calculation or from temperature measurements. It has sometimes been stated that the solidus temperature for the alloy determines the end of solidification; however, while this is true in the very special case of the advance of a planar interface under steady-state conditions (see page 132), it is not true under the more practical conditions of a cellular-dendritic interface; the end of solidification must occur at the end of the "terminal transient" for the interdendritic liquid. If diffusion were sufficiently fast for equilibrium to be maintained, then the process would terminate at the solidus temperature, but this would require diffusion in the solid as well as in the liquid, and it is only with interstitial solutes that this can occur at significant rates. In other cases there is certain to be interdendritic segregation and liquid will be present down to a limiting temperature such as that of a eutectic. In principle, the only ways to avoid dendritic or cellular-dendritic growth are to decrease the speed and to increase the temperature gradient, but it is most unlikely that it is practical to pour a continuous casting with sufficient superheat and to solidify at a slow enough rate to avoid dendritic solidification and its attendant complications.

FLUID MOTION IN CONTINUOUS CASTING. In continuous casting of metal with relatively high thermal conductivity, such as aluminum, the liquid pool can be quite shallow; and it is then found that the way in which the liquid metal enters the mold has an important effect on the structure. If the liquid stream sweeps over the surface, a very fine grain structure is produced, while more quiescent pouring, in which the stream is slowed down by the pool so that it does not sweep over the surface, a completely columnar structure may be produced.

Effect of vibration on structure. Many investigators have found that grain refinement can be brought about by introducing vibration while the metal is solidifying (28–40). There seems to be general agreement that vibration is effective only if it is imposed while the metal is solidifying, and not while it is superheated. There appear to be two distinct views as to the mechanism; one is that the vibration actually causes more nucleation to occur than would otherwise be the case. The process in this case would, presumably, be of the cavitation type discussed in Chapter 3, and the implication would be that a particle that did not nucleate under static conditions could be made to do so by cavitation at its surface. The other view is that vibration has much the same effect as turbulence, dispersing small crystals so that more of them grow with the result that the grain size is reduced. There does not at present appear to be any decisive way of discriminating between these proposals. In addition to the grain-refining effect it is reported (36) that degassing is improved by vibration, with the result that porosity is reduced; on the other hand, Richards and Rostoker (32) find no effect on microporosity or segregation, and report that too much vibration can cause hot tearing.

One of the few quantitative investigations on the effects of vibration is that of Richards and Rostoker (41), who studied the effect of the frequency and intensity of vibration on the grain size in an aluminum alloy with 4½ per cent copper. Their experiments were conducted on 50-pound ingots and the vibration was at 60, 300, and 1500 cycles. They found that the grain size was decreased by increasing intensity of vibration, as shown in Fig. 8.25, in which it will be seen that the intensity of vibration is represented by the amplitude. It is interesting that the amplitude is apparently a more significant measure of vibration than either the maximum velocity, the acceleration or the energy; this follows from the result that if the grain size is plotted against amplitude, as in Fig. 8.25, then the relationship is independent of frequency, whereas it becomes frequency dependent if the other parameters are used. Another interesting observation was that the pouring temperature has no effect on the grain size if vibration was

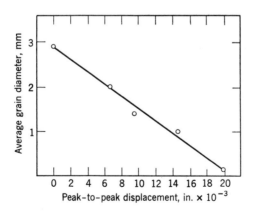

Fig. 8.25. Effect of amplitude of vibration on grain size. (From Ref. 41, p. 884.)

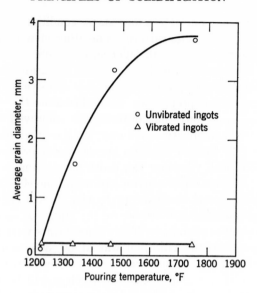

Fig. 8.26. Effect of pouring temperature on grain size. (From Ref. 41, p. 884.)

applied during solidification, but that the usual effect was observed in the absence of vibration (Fig. 8.26). In their discussion of their results, Richards and Rostoker point out that not all alloys show grain refinement; in fact, in alloys in which a nondendritic phase is formed (as in aluminum alloys containing silicon or iron, and cast iron) the effect of vibration is apparently to coarsen the structure. This is regarded as an indication that the effect of vibration cannot be, as has been suggested, the detachment of fragments of growing crystals. However, this deduction does not appear to be conclusive, because the possibility of detachment may well depend critically on the morphology of the interface. Richards and Rostoker suggest that the effect is due to an increase in nucleation rate which, they point out, could be caused by pressure changes. Their theory is too qualitative and speculative to be subjected to any useful comparison with experiment.

Two more recent papers by Tiller and his coworkers (42 and 43) describe experiments which demonstrate the importance of good coupling between the vibrator and the melt, and indicate how this can be achieved. The first of these papers contains a discussion of the theory, in which five distinct hypotheses are examined; these are as follows:

1. That the vibration changes the distribution of solute, which changes the equilibrium temperatures, locally increasing the constitutional supercooling. This is rejected on the basis that extra mixing would necessarily reduce the constitutional supercooling.

2. That vibration changes the actual temperature distribution in the melt. This is rejected on the basis that any extra energy introduced into the melt (by attenuation of the vibration) would decrease the rate of solidification; in any case, the amount of energy used is too small to have a significant direct effect.

3. The vibration has a direct effect on the nucleation rate (this is equivalent to the view advanced by Richards and Rostoker). Tiller et al. suggest that this could be caused by (a) local temperature inhomogeneities from stirring action, (b) "cleaning" the nucleant particles by eroding them, (c) agglomeration of subcritical nuclei, (d) removal of fragments of the solid (this is however not equivalent to increasing the nucleation rate), or (e) a direct effect of pressure on the solid-liquid equilibrium conditions as proposed by Walker and discussed in Chapter 3. Tiller concludes that the effect of pressure, as amplified by cavitation, is the most probable explanation of the phenomena.

The question is still far from settled, however, since there are no predictions sufficiently precise for quantitative comparison with experiment. From a practical point of view, it should be noted that all the successful experiments on the use of vibration for grain refinement have been made on rather small samples and the problem of transmitting useful amounts of vibration into the interior of large solidifying masses has not yet been solved.

Welding, brazing, and soldering as casting processes. A very important and widely used type of casting operation is that in which the space between two pieces of metal is filled with a molten metal, which is then allowed to solidify. The process, called welding, brazing, or soldering according to the choice of filler, is successful if the "casting" adheres properly to both pieces and is mechanically strong enough. There are, therefore, two main problems that are directly related to solidification—adhesion and strength. The adhesion of welds, in which the filler is similar in composition to the metal being joined, depends upon the establishment of continuity of structure; that is to say, the weld metal should begin to solidify as a columnar structure formed by the growth of crystals of the previously solid metal. In order to achieve the necessary "wetting" of the solid metal by the molten metal it is necessary to melt at least a thin layer at the surface. This forms a liquid layer "behind" the oxide or other layers on the surface; the oxide is no longer supported by solid metal and is quickly broken up or pushed aside as the two liquids make contact with each other. In the case of soldering of steel with a tin-lead solder, adhesion appears to arise from the formation of an intermediate layer consisting of the intermetallic phases of the tin-iron system. The formation of this layer, which occurs in the "tinning" part of the soldering operation, requires a sufficiently high temperature and the use of a flux which either dissolves the oxide on the steel surface (active flux) or at least prevents its re-formation or growth (passive flux). "Brazing" includes cases that are basically similar to welding, from this point of view, although the procedures are different; for example, brazing of aluminum alloys with the aluminum silicon eutectic alloy. Some types of brazing are similar to soldering except for the higher temperature that is required and greater strength that can be produced. The brazing of steel with copper base brazing alloys is an example.

The strength of the weld depends on the structure that is produced during solidification and on subsequent damage of the type mentioned on page 251. The solidification process in a weld differs from those discussed above only in the nature of the "mold," the initial temperature distribution and the size of the "casting," and will therefore not be discussed further here.

8.3   Segregation

Segregation may be defined as describing all departures from uniform distribution of the chemical elements that are present. The segregation that is actually observed in an alloy or an impure metal is the result of the rejection of solute at the interface followed by its redistribution by diffusion and by mass flow. If the redistribution were by diffusion alone, or if the conditions were sufficiently well controlled for the value of the effective distribution coefficient to be predictable, then the final distribution of solute could be calculated from the geometry of the interface and its rate of advance. The only case in which this is possible is that of zone refining. In all other cases of practical interest, the complexity of the shape of the interface and uncertainty about the actual mass flow in the liquid conspire to make useful quantitative predictions impossible. The complexity of the shape of the interface arises from the instability, under most real conditions, of a planar interface and its breakdown into cellular, cellular-dendritic or dendritic forms. The liquid that is enriched with rejected solute is caused to move by five separate effects: (1) the motion of the liquid as it enters the mold, (2) convection caused by differences of density due to temperature differences, and (3) convection caused by differences of density due to variations in composition of the liquid as a result of rejection of solute, (4) motion, caused by gravity, of crystals that are growing in the liquid, and (5) motion of liquid due to solidification shrinkage.

A detailed analysis of segregation in real systems is, therefore, impossible, but the following generalizations are useful. It is desirable to distinguish between short-range or micro-segregation and long-range or macro-segregation in castings and ingots. The characteristic distance which should be used for making this distinction is the distance between neighboring "walls" in the cellular-dendritic substructures; if there is a significant variation of composition within the microstructural units defined by the substructure, there is short-range segregation; if the composition, averaged over several substructural units, varies from one region of the cast metal to another, then the segregation is of long range. It has been shown in special cases (44), and it is probably always true, that significant long-range segregation is caused by mass movement of the enriched liquid and not by diffusion alone. Thus normal segregation, inverse segregation, and gravity segregation are associated with motion of the liquid, and are completely unpredictable on the basis of diffusion alone. It has been shown (see page 181) that the mass flow required for inverse segregation is sufficiently localized to be unaffected by pouring turbulence, or convection, and is therefore predictable; but normal segregation and gravity segregation are the results of motion over distances comparable with the size of the mold, and are therefore not readily predicted theoretically and cannot be studied quantitatively by means of small-scale experiments.

The most important type of short range segregation in ingots and castings is that which occurs within the cells of the cellular-dendritic structure. It is shown on page 178 that predictions based on an idealized geometrical model are well substantiated by experiment, and it follows that the assumptions underlying this analysis are valid. We may, therefore, conclude that, while the extent of normal segregation and gravity segregation cannot be predicted without detailed knowledge of the fluid motion that takes place during solidification, micro-segregation and inverse segregation can, in principle, be predicted quantitatively; the effectiveness of measures that might be proposed for reducing or eliminating either of these types of inhomogeneity can be studied by means of small-scale experiments, as well as theoretically, with confidence that the phenomena are well understood.

8.4 The Significance of Small-Scale Experiments

It would be very useful to be able to predict the structure and properties of a full scale ingot or casting from the results of small-scale experiments; it has sometimes been stated that it should be possible to scale down the whole of a casting operation so that the result would show geometrical similarity with that of the full scale process. This would require the existence of nondimensional parameters, such as are found, for example, in hydrodynamics, in the Reynolds number. It may easily be seen, however, that, even in the simplest (but unrealistic) case of a casting in which the liquid is stationary, the criteria which determine grain size are related to the potency of the nucleant, and to the rates of fall of temperature near the interface and in the interior of the liquid. The heat capacity of the casting increases with the cube of its linear dimensions, whereas the surface area increases with the square, and the distance over which heat must be conducted increases as the first power; and therefore the time scale for the cooling increases in proportion to the square of the linear dimensions. That is, all processes that depend only on rates of cooling will be affected in the same way; but the supercooling required for nucleation does not depend on the size of the casting, and so the number of nuclei formed will be neither constant nor proportional to the size of the casting. The constitutional supercooling effects, on the other hand, depend on the time scale for cooling (i.e., rate of advance of interface, R) and on the temperature gradient G, which decreases linearly with the linear size of the casting if the center and surface temperatures are fixed.

The criteria for cell formation depend upon G/R, for cellular-dendritic growth upon $G/R^{1/2}$, and for equiaxed structure upon G, R, and the nucleation catalyst, so that the heat extraction of the mold cannot possibly be controlled, by choice of mold dimensions, so that all stages of the process occur at relatively the same position in the mold, that is, at the same fraction of the distance from wall to center.

The best that can be hoped for is to select a mold wall thickness that allows one of the criteria to be scaled, and to determine separately the effectiveness of the nucleants that are present. But the usefulness of these procedures is severely limited by the large, but not precisely known, effect of fluid motion on the grain size, on the length of the columnar zone, and on the extent and distribution of long-range segregation.

8.5   Change of Volume on Freezing

All the metals which are used, either in pure form or as the basis for alloys, on account of their mechanical properties, have close-packed or nearly close-packed structures; that is, face-centered cubic, close-packed hexagonal or body-centered cubic. The liquid form of any metal has a density lower than that of the solid with any of these crystal structures, and consequently, all these metals and their alloys contract on solidification. Table 8.1 indicates the extent of this con-traction. With the exception of a few alloys used for special purposes, all alloys contract on solidification; the implications of this will now be discussed.

Table 8.1.   Change of Volume on Solidification

Element	ΔV (per cent)
Aluminum	−6.0
Magnesium	−5.1
Cadmium	−4.7
Zinc	−4.2
Copper	−4.1
Silver	−3.8
Mercury	−3.7
Lead	−3.5
Tin	−2.8
Sodium	−2.5
Potassium	−2.5
Iron	−2.2
Lithium	−1.65
Antimony	+0.95
Gallium	+3.2
Bismuth	+3.3

The most obvious consequence of shrinkage on solidification is that a quantity of metal that just fills a mold cavity when it is molten will no longer do so after it is solid. This cannot be prevented by the application of force, because the negative pressure that would be re-quired to produce, for example, 5 per cent elastic expansion would be far greater than the metal would withstand. Consequently, it is the location, rather than the presence, of shrinkage that should be con-sidered. There are five main ways in which shrinkage can manifest itself, represented diagrammatically in Fig. 8.27. These may be de-

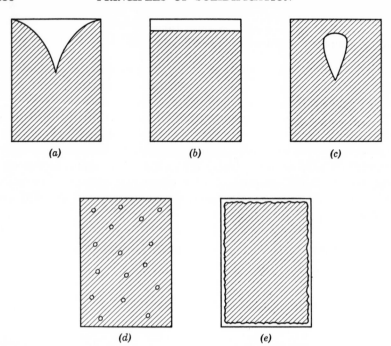

Fig. 8.27. Shrinkage effects. (a) Pipe, (b) unidirectional shrinkage, (c) cavity, (d) distributed porosity, (e) surface porosity.

scribed as (a) pipe, (b) unidirectional shrinkage, (c) cavity, (d) distributed porosity, (e) surface porosity. The total shrinkage may take one of these forms or it may be divided between more than one of them. The mode of solidification and the direction of solidification jointly determine the position of the shrinkage; the conditions which give rise to each of the five types will now be discussed.

Pipe formation. The shape of the pipe which is formed depends upon the relative extent of solidification from the sides and from the bottom. If the shape and the thermal environment are such that cooling is effectively only from the sides, then the shape would be that shown in Fig. 8.28a, in which the final form and an intermediate stage are shown. The shape is a result of the fact that as solidification proceeds, the formation of a given volume of solid reduces the volume of the remaining liquid by an amount that remains constant, but by a proportion that increases as the quantity of remaining liquid decreases. If solidification is mainly from the bottom, the shape of pipe is that shown in Fig. 8.28b.

Unidirectional shrinkage. The depth of the pipe decreases as the relative amount of cooling from below increases, until, when cooling is only from below, the final surface will be flat, at a distance below the original surface that depends only on the initial depth and the volume contraction. This is an example of unidirectional freezing. It should be remembered that uniform shrinkage occurs during the subsequent cooling of the metal after it is solid; this may cause a change of dimensions comparable with that which occurs *during* solidification. For example, a low carbon steel has an average coefficient of thermal expansion of about $1.5 \times 10^{-5}/°C$ between room temperature and the melting point (1500°C). The change of volume *after* solidification therefore is about 6½ per cent.

Cavity formation. In many practical cases, there is sufficient heat loss from the upper, or free, surface of the liquid for a solid "crust" to form on the top at the same time that it forms along the sides and bottom. In this case, the last region to solidify is in the interior, where its shape would ideally be as shown in Fig. 8.27c. It often happens that the crust is not strong enough to withstand the pressure of the atmosphere above it when the cavity forms below; this may lead to partial collapse of the top crust into the cavity.

Control of unsoundness. The three effects of shrinkage discussed so far are macroscopic in extent, and their occurrence cannot be prevented, because the shrinkage that occurs on solidification is an inherent part of the process. On the other hand, the presence of a pipe or a cavity in a casting may be highly detrimental to its properties. The gross effects of shrinkage are always to be found where solidification occurred last, and their *position* (but not their existence) depends upon the relative amounts of heat extracted by different regions of

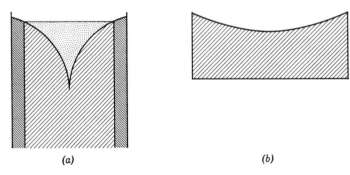

(a)   (b)

· Fig. 8.28. Geometry of pipe. (a) Mainly lateral cooling, (b) mainly vertical cooling.

the mold.   While it would not be appropriate to discuss detailed techniques, it may be pointed out that there are three ways in which the direction or rate of heat flow can be modified; these are by the use of *chills,* which are inserts of high thermal conductivity in a mold of lower conductivity, by means of *padding,* i.e., the local retardation of heat extraction by the use of material of relatively low thermal conductivity or by local preheating, and by the use of *exothermic compounds,* that is, mold material or dressing that gives out heat and thereby retards the cooling of the adjacent metal.   It is evident that these methods can supplement, but not replace, attention to the geometry of the casting, since the most important parameter that controls the cooling of a given region is its cross section, that is, the ratio of its thermal capacity to the area through which heat is extracted.   The most desirable geometry is that which produces a heat flow pattern in which the parts that are to be rejected (e.g., risers) are the last to solidify and in which no liquid is ever isolated from the supply provided by the risers.

Porosity.   The three shrinkage effects discussed above are macroscopic in scale; we must next consider the consequences of the fact that most practical materials are cellular-dendritic or dendritic in structure.   One of the consequences of cellular dendritic solidification is that the liquid is subdivided into thin "slices" that solidify essentially transversely.   A typical inter-wall distance is 0.01 mm, while the depth may be 1 cm or more; this is represented in Fig. 8.29.   It has been shown by Flemings (45) that the liquid at *A* has too high a viscosity to be "sucked" into the groove sufficiently fast to keep pace with the shrinkage that accompanies solidification on the walls of the

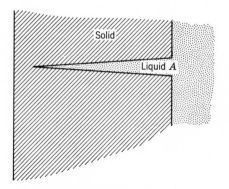

Fig. 8.29.   Origin of microporosity.

groove; pores, therefore, are usually formed. These are "micropores," the presence of which is, it is believed, unavoidable in alloys that solidify with the typical cellular-dendritic structure unless the structure is extremely coarse. It should be noted that the liquid in which the pore is forming is continuous with the bulk liquid, and therefore "feeding" would be possible if sufficient time were available for the liquid to flow to the required extent. The size and number of the pores that are formed depend upon many factors; an important one is the extent to which gas tends to come out of solution (see p. 187). If the supersaturation of gas is substantial, it will be easier for pores to nucleate than if the supersaturation is less. Thus a high gas content is conducive to the formation of a large number of very fine pores (microporosity) while a low gas content encourages the production of fewer, much larger, pores. The rate of cooling also has some influence on the size of the pores, because this affects the distance through which gas can diffuse during the period when pores are nucleating and growing. It is not realistic to distinguish between "shrinkage porosity" and "gas porosity," as is sometimes done, as if they were two entirely different kinds of defect. It is, however, to be remembered that if there is gas evolution in the metal or at the mold wall, blowholes can be formed which are not associated with shrinkage. These effects are discussed on page 191.

Surface porosity. The origin of surface porosity is discussed on page 294 in connection with the factors that govern surface topography.

8.6 Blowholes

Gas may be present in a solidifying metal or alloy for several reasons, of which the most important is a chemical reaction such as the oxidation of carbon to form carbon monoxide or dioxide; this occurs, for example, in a rimmed steel, in which the evolution of gas during the early stages of solidification is sufficient to keep the liquid near the solid-liquid interface thoroughly mixed with the bulk liquid. The first part to solidify is, therefore, low in carbon and relatively free from other solutes. The last part to solidify usually contains many trapped gas bubbles. If, on the other hand, a melt of steel contains excessive hydrogen, the rejection of hydrogen on solidification may be sufficient to cause the solidifying ingot to "boil." This is not to be interpreted as actual boiling of the liquid steel; it is the effect of the evolution of large numbers of hydrogen bubbles. Most escape at the surface, but some are trapped, resulting in a "wild" or porous ingot. The hydrogen content in such cases is 8–12 cubic centimeters per 100 grams (46). Simialowski (47) gives the following upper limits for hydrogen for the production of sound ingots:

Carbon steel	6.5 cm^3/100 gm
Low alloy steel	7 cm^3/100 gm
Chromium (ferritic) stainless	10 cm^3/100 gm
Austenitic stainless	12 cm^3/100 gm

In this case the evolution of gas takes place without a chemical reaction. If the gas content, or more precisely the amount of gas that is rejected during solidification, is insufficient to cause "boiling," there may still be undesirable consequences; the gas may interact with shrinkage to cause porosity; or if there is no negative pressure due to shrinkage, the gas may still evolve in the form of bubbles, which may either detach themselves and float up to the surface, be trapped as blowholes, or grow as cylindrical cavities, analogous to the cylindrical bubbles often seen in ice cubes. The condition for the growth of cylindrical bubbles is that gas should diffuse to the bubble from the surrounding melt at a rate that keeps up with the advance of the interface (see Fig. 6.6).

8.7 Surface Topography Resulting from Solidification

One of the many criteria for the success of a solidification process is the topography of the resulting surface. The surface should, for most purposes, match the mold surface as precisely as possible; or, in regions in which the metal is not in contact with a mold, it should approach geometrical smoothness as closely as possible. The importance of this criterion, in comparison with others (including cost), varies greatly from one case to another; for example, type for printing or a precision casting to be used without any machining or other finishing must reproduce the required shape to within very close tolerances, while an ingot that is to be rolled to sheet, on the other hand, is not required to match the mold shape nearly so precisely. A superficial view of this question would indicate that when a metal solidifies in contact with a mold surface through which heat is extracted, the metal would match the mold shape exactly; exactly, in this context, would mean that the shape is limited only by the atomic nature of matter. It is very evident, however, that the precision is not nearly as good as this would imply. It is extremely difficult, if not impossible, to produce an as-cast surface that is as smooth optically as the mold surface; this means that the departure from exact replication is at least 10^{-5} cm. In most cases, the as-cast surface departs from the topography of the mold by amounts that would be measured in millimeters.

Many causes have been identified for the difference in shape between a metal surface and the mold surface, and for the often undesirable shapes taken up by free surfaces of solidifying metals. Although they are not mutually exclusive, it is convenient to discuss them separately here.

Surface tension effects. It is well known that one of the manifestations of the surface tension of liquids is that a pressure difference equal to $P = \sigma(1/R_1 + 1/R_2)$ exists between the inside of a liquid and the adjacent medium if the surface has principal radii of curvature R_1 and R_2 and surface tension σ. A consequence is that a mold with sharp edges or corners can be filled only to the extent permitted by the pressure in the liquid. In general, metals appear to have a surface tension of about 500 dynes/cm; the pressures required to fill a corner to various radii of curvature are shown in Fig. 8.30, from which it will be seen that a radius of 1/100 mm (i.e., 10^{-3} cm) can be achieved with an excess pressure in the liquid of about 1 atmosphere. The excess pressure can be the result of a hydrostatic "head" of liquid, or it can be provided as the centrifugal force in a rotating liquid, or it can be applied directly as in a pressure die casting; alternatively, part of the pressure difference (up to one atmosphere) can be achieved

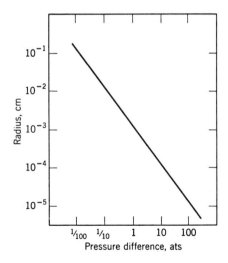

Fig. 8.30. Limiting radius of curvature of surface of a liquid as a function of pressure difference.

by exhausting the gas out of the mold through the pores of the mold itself. It is evident that the hydrostatic head of the liquid metal in the mold and its risers is usually very limited; liquid steel has a density of about seven, and therefore a pressure difference of one atmosphere would require a head of about one and a half meters, or four and a half feet. In a small casting, in which sharp corners may be desirable, the head is not likely to exceed one tenth of this value, giving a minimum radius of about a tenth of a millimeter unless extra pressure is used.

Cold shut. A cold shut is formed when some liquid metal solidifies in contact with, but not continuously with, a previously solidified surface. The failure to unite to form continuous metal is usually due to the presence of an oxide film, either on the liquid surface or on the solid. This type of defect is often caused by metal splashing as it enters the mold, with the result that drops or splashes make contact with the mold above the general liquid metal level. These drops may be chilled by contact with the wall and become solid before the main liquid level reaches them. If the temperature of the liquid being poured is not sufficiently high to melt these solid regions, cold shuts may be formed. This type of defect can also be caused by allowing a metal to enter a mold at more than one place, in such a way that liquid impinges on solid metal.

Trapped gas. Solidification may begin by the formation of a continuous skin of solid; contact between this skin and the mold wall may be prevented locally by gas that was trapped in interstices in the mold material, or by water vapor, coming from mold or mold-dressing. Even if the latter is porous, there may not be sufficient time for the trapped gas to escape completely, and it must be remembered that the gas expands as it heats up from its original temperature (that of the mold) to the temperature of the hot metal. Contact may also be prevented by gas bubbles resulting from a mold-metal reaction.

Surface dendrites. It has already been pointed out that the structure at and near the surface may either be that of the crystals that are formed during the rapid chilling that may occur when the metal enters the mold, or it may be a structure that is formed after the original chill structure has been remelted by the hot metal. In the former case, the crystals are small, and grow very fast in directions parallel to the surface. The surface layer may be formed dendritically, but the interdendritic spaces may not be filled completely; if large crystals are formed after considerable supercooling has occurred, the type of relief shown in Fig. 8.31a may be found. On the other hand,

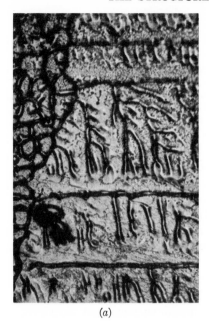

(a)

Fig. 8.31. Surface dendrites. (a)
Supercooled, ×100; (b) slowly cooled,
×1). (Photographs by (a) D. Maisel,
(b) R. B. Williamson.)

(b)

when remelt has occurred, the subsequent cooling is much slower, and
fewer crystals are formed in contact with the wall. There is far less
supercooling while the crystals are growing along or close to the mold
surface; in an alloy, however, there is certain to be constitutional
supercooling in the liquid close to the wall. Growth is therefore again
dendritic; but it is much coarser than in the chilled crystals because
the supercooling is much less. The interdendritic spaces eventually
fill in, but this must occur largely by thickening of the dendrite arms,
and the corresponding shrinkage can therefore pull liquid away from
the mold wall. The dendritic pattern will, in these circumstances,
stand out in relief on the as-cast surface. Examples are given in Fig.
8.31. Dendritic relief is not usually found unless either (a) the melt
was supercooled considerably (as in Fig. 8.31a) or remelting had

occurred (Fig. 8.31b). It is necessary for the crystal to be relatively large, because otherwise the structure is entirely pre-dendritic.

Exudation and surface porosity. More drastic consequences of melt-back occur when the dendritic skeleton formed during refreezing becomes strong enough to shrink away from the mold before the interdendritic liquid has solidified. The thermal contraction of the dendritic shell compresses the liquid, and the space between shell and mold applies suction to the liquid, which may be forced outward to form drops on the surface. These drops, being composed of inter-dendritic liquid, are much richer in solute than the bulk composition of the shell, and they therefore represent a form of inverse segregation. If the liquid in the mold freezes over early in the process, "negative pressure" or suction may develop. In this case interdendritic liquid may be sucked inward, leaving severe interdendritic porosity at and near the surface. This type of porosity is sometimes attributed to metal-mold reactions.

Topography of the free surface. The topography of the free surface of a solidifying metal is usually controlled by surface tension, which determines the *shape* of the liquid surface, and shrinkage, which causes the *level* of the surface to change as solidification proceeds. Special conditions are necessary to avoid shrinkage effects; but even if this is achieved, there is still another mechanism by which the surface departs from optical perfection. This is the formation of "terraces" on the surface of the solidifying metal. These are regions of the surface parallel to specific crystallographic planes; they are formed because the most closely packed atomic planes have the lowest specific surface energy. The exact mechanism of the formation of the terraces has not been identified, and it is not known how general the phenomenon is. It occurs in lead (48) and germanium and may be much more general.

Changes of surface topography after solidification. There are three mechanisms by which the surface of an as-cast metal can change after solidification is complete: (1) chemical reaction, such as the formation of oxide, (2) thermal etching, which can cause grooves at the grain-boundaries and terracing of the surfaces, and (3) vacancy pit formation (49). When a metal cools, its equilibrium vacancy content decreases. If sufficient time is allowed, the excess vacancies diffuse to the surface and form pits, usually pyramidal in shape. These pits are small and close together and give the surface an etched appearance. Their absence close to other vacancy sinks, such as grain boundaries, sometimes reveals the presence and position of such structural features.

Metal coating by hot dipping. There are several industrial processes in which a metal is coated with a lower melting point metal by dipping it into a molten bath. Tinplate and galvanized iron are examples; and although these processes are now often performed by electrolysis, which is outside the scope of this book, the traditional hot dipping processes are still of interest. In hot dipping, the protective and decorative value of the coating depends upon its continuity and its topography, and, in the case of hot dipped zinc, on the size and shape of the crystals.

Tinplate. Hot dipped tinplate is made by immersing suitably prepared sheets or strips of steel in molten tin, and withdrawing it through rollers that limit the amount of tin remaining on the surfaces. As a result of the well-known effect of "surface cavitation" the tin is distributed into a "river structure" as it leaves the roller. This event is due to the fact that as the roller separates from the sheet, the pressure in the liquid decreases, so that it develops an instability that is relieved by most of the liquid dividing into distinct streams. An example of the "river" structure is shown in Fig. 8.32. This surface topography is frozen in as the tin freezes very soon after leaving the rolls.

Other topographical effects on the surface of tin are caused (50) by the tendency to reach local surface tension equilibrium either with spots on the substrate that are not wetted (Fig. 8.33a) or with drops of flux that rest on the surface (Fig. 8.33b).

Hot Dipped Zinc. Galvanized iron often has a distinctive surface appearance called "spangle." In this structure, each crystal is clearly

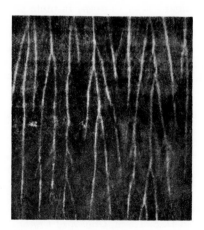

Fig. 8.32.   River structure on tinplate.

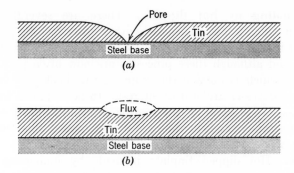

Fig. 8.33.   Surface defects on tinplate.

visible, as a result of slight etching that takes place during cooling. The presence of spangle indicates that nucleation of crystals was relatively infrequent and that considerable undercooling must occur.

References

1. B. Chalmers, *J. Australian Inst. Met.,* **8,** 255 (1963).
2. D. Walton and B. Chalmers, *Trans. Met. Soc. AIME,* **215,** 447 (1954).
3. C. S. Smith, *Trans. Met. Soc. AIME,* **175,** 15 (1948).
4. B. Chalmers, *Proc. Roy. Soc., Ser. A,* **196,** 64 (1949).
5. H. Biloni and B. Chalmers, Unpublished work.
6. W. C. Winegard and B. Chalmers, *Trans. Am. Soc. Met.,* **46,** 1214 (1954).
7. J. L. Walker, *Private Communication.*
8. W. A. Tiller, *Trans. Met. Soc. AIME,* **224,** 446 (1962).
9. B. Chalmers, *Physical Metallurgy,* John Wiley and Sons, New York, 1959.
10. R. W. Ruddle, *Solidification of Castings,* Institute of Metals, London, 1957, p. 59.
11. W. Roth and M. Schippen, *Z. Metallk.,* **47,** 78 (1956).
12. F. A. Crosley, R. D. Fisher, and A. G. Metcalfe, *Trans. Met. Soc. AIME,* **221,** 419 (1961).
13. F. C. Langenberg, G. Pestel, and C. R. Honeycutt, *Trans. Met. Soc. AIME,* **221,** 993 (1961).
14. H. F. Bishop and W. S. Pellini, *Foundry,* **80,** 86 (1952).
15. R. W. Ruddle, *Solidification of Casting,* Institute of Metals, London, 1957.
16. W. Thurz, *Metall.,* **9,** 580 (1955).
17. S. Terai, *Nippon Kinzoku Gakkaishi,* **19,** 249 (1955).
18. H. Bernstein, *Trans. AIME,* **200,** 603 (1954).
19. H. G. Warrington, *Progr. in Met. Phys.,* **2,** 121 (1950).
20. W. A. Baker, M. D. Eborall, and A. Cibula, *Inst. Met.,* **81,** 43 (1952–1953).
21. A. Schneider and G. Strauss, *Aluminum,* **37,** 712 (1961).
22. J. P. Dennison and E. V. Tull, *J. Inst. Met.,* **24,** 858 (1956–1957).
23. N. L. Pokrovsky and D. S. Tissen, *Growth of Crystals,* **3,** 150 (1962).
24. *Metals Handbook,* American Society for Metals, Cleveland, 1948, p. 959.

25. R. D. Green, *Trans. Am. Foundrymens' Soc.*, **66**, 380 (1958).

26. M. C. Flemings, *Private Communication.*

27. V. I. Dobatkus, *Metallurgy*, **14**, 101 (1948).

28. J. B. Jones, *USAEC Publ. AF-TR 6675*, 1951.

29. W. Rostoker and M. J. Berger, *Foundry*, **81**, 100 (1953).

30. R. G. O'Rourke, *USAEC Publ., AECU 2535*, 1953.

31. W. M. Govorkov and K. N. Shabalin, *Zh. Tekhu Fiz.* (1956).

32. R. S. Richards and W. Rostoker, *Trans. Am. Soc. Met.*, **48**, 23 (1955).

33. G. P. Kuslita and B. G. Strongin, *Fiz. Metallov i Metallovednie*, **5(1)**, 187 (1957).

34. V. A. Serkovsky, *Liteinoe Prorzvodstio*, **5**, 19 (1958).

35. N. N. Sinota, E. A. Lekhtblan, and E. M. Smolyavenko, *POMM*, **7**, 77 (1959).

36. M. B. Al'tman, V. I. Slotin, N. P. Stromskaya, and B. I. Eskin, *Izv. Akad. Nauk SSSR*, Met. 1959 (Tekhu), **3**, 88.

37. F. Erdmann-Jesnitzer and J. Schubert, *Nene Hütte*, **4**, 359 (1949).

38. Y. Hari and I. Vezawa, *Nippon Kinzoku Gakkishi*, **23**, 168 (1959).

39. D. H. Lane, J. W. Cunningham, and W. A. Tiller, *Trans. Met. Soc. AIME*, **218**, 985 (1960).

40. G. I. Levin and I. G. Polotsky, *Izv. Akad. Nauk SSSR Met. i Tope* (Tekhu), **5**, 167 (1961).

41. R. S. Richards and W. Rostoker, *Trans. Am. Soc. Met.*, **48**, 884 (1956).

42. D. H. Lane, J. W. Cunningham, and W. A. Tiller, *Trans. Met. Soc. AIME*, **218**, 985 (1960).

43. D. H. Lane and W. A. Tiller, *Trans. Met. Soc. AIME*, **218**, 991 (1960).

44. F. Weinberg, *Trans. Met. Soc. AIME*, **221**, 844 (1961).

45. M. C. Flemings, *Private Communication.*

46. C. Sykes, H. H. Burton, and C. G. Gegg, *J.I.S.I.*, **156**, 155 (1947).

47. M. Smialowski, *Hydrogen in Steel*, Addison Wesley, Reading, Mass., 1962, p. 278.

48. H. A. Atwater, A. R. Lang, and B. Chalmers, *Can. J. Phys.*, **33**, 352 (1955).

49. P. E. Doherty and R. S. Davis, *Acta Met.*, **7**, 118 (1959).

50. B. Chalmers, *Trans. Faraday Soc.*, **33**, 1167 (1937).

Appendix

The Production
of Single Crystals
from the Melt

A.1   General Considerations

It has been realized increasingly during the last thirty years that many of the properties of solids can be studied most efficiently in single crystal specimens; this led to considerable research in the theory and practice of single crystal growth. Much of the knowledge that came from these investigations has been applied to the production of crystals for an ever widening range of "Solid State Devices." Many of the crystals that are used for research and for applications are prepared by the controlled solidification of a melt or of part of a melt; there are, of course, other methods, such as recrystallization of the solid and condensation of the vapor, that are widely used, but these lie outside the scope of this book and will not be discussed here.

All the methods that are used for growing crystals from the melt are based on the simple principle that the extraction of latent heat must be achieved without allowing the melt to supercool sufficiently to permit nucleation of new crystals; this almost always requires that the extraction of heat must be by conduction through the existing crystal. In practice this requires a heat sink that removes heat from the crystal, and a heat source that supplies heat to the melt. Any system that maintains in the melt a temperature gradient that is sufficient to prevent nucleation, and removes heat through the crystal, will permit the growth of an existing crystal. The problems that arise are a result of the fact that it is nearly always necessary to control one or more of the following characteristics of the crystals that are to be grown: orientation, shape, composition, perfection. Each of these aspects of crystal growth will be discussed in turn.

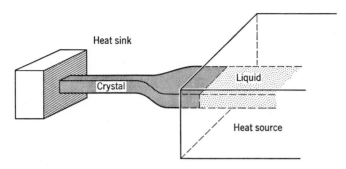

Fig. A.1. Use of seed crystal to control orientation.

A.2 Control of Orientation

For many purposes it is either necessary, or at least convenient, to produce crystals of specified orientations. This is achieved by the use of a seed crystal of which one end is in contact with the heat sink, while the other end is melted by contact with the liquid. This is shown schematically in Fig. A.1 in which the problem of containing the melt has been ignored; it is discussed below. Growth of the crystal is by movement to the right of the interface between seed and melt, and the orientation of the resulting crystal will, within limits to be discussed below, be identical with that of the seed. The main problem that is inherent in this method is the production of the original seed crystal of the required orientation. This can be achieved as follows: If a mass of metal, such as that shown in Fig. A.2, is completely melted, and is then allowed to cool, it will nucleate at a large number of points and will consist of many crystals. If it is allowed to cool with a temperature gradient, so that, for example, the end A is always the coldest part of the specimen, then several crystals may nucleate when A reaches the nucleation temperature, but no more crystals will form thereafter because the initial crystals can grow and prevent the recurrence of supercooling. Under these conditions it is sometimes found that one of the original crystals gradually achieves dominance by suppressing its neighbors, in which case the

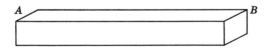

Fig. A.2. Polycrystalline casting for conversion into a single crystal.

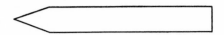

Fig. A.3. Optimum shape for growth of single crystal without a seed.

major part of the specimen would be a single crystal. This can be made more probable (but not certain) by reducing the cross section of the region where solidification begins so as to increase the chance that only one crystal will be formed, or will survive (Fig. A.3).

The crystal formed by this process is unlikely to have the required orientation; however, a crystal with a *selected* orientation can be grown from one of known orientation by orienting the seed appropriately, as shown in Fig. A.4, in which it is assumed that a crystal is to be grown with an ⟨001⟩ axis in the longitudinal direction. The seed has an ⟨001⟩ axis at about 40° to its geometrical axis. If seeding is successful, the ⟨001⟩ axis of the crystal will be in the same direction as that of the seed, as shown in the diagram.

It is sometimes convenient to produce a crystal of approximately the required orientation, by a first seeding operation, and to correct it by means of a second operation. By this method, a seed crystal can be obtained with any desired orientation, although a change of more than about 40° usually requires two stages. If a goniometer mount is used for adjusting the orientation of the seed, any required precision can be achieved. The question of whether an initial orientation is precisely maintained during growth is discussed under the heading of "Control of Perfection."

A.3   Control of Shape

There are four main methods that are used for the growth of single crystals from the melt; of these, two make use of a mold to contain

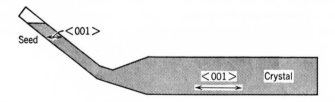

Fig. A.4. Crystal of required orientation from seed of a different, but known orientation.

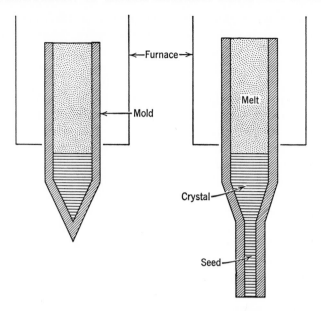

Fig. A.5. Bridgman method for growth of single crystals.

the metal; one is the Bridgman method, in which a mold with its long axis vertical is slowly lowered through a furnace, so that solidification begins either at a point at the bottom or on a seed that is in a downward extension on the mold (Fig. A.5).

A horizontal "boat," shown in Fig. A.6, is used in the alternative method. The seed is placed in the space S, and the melt fills the region M. Solidification proceeds from left to right as a furnace is moved to the right. Heat loss from the boat at A is often a sufficient heat sink, but a water cooling tube may be embedded in the boat. If specimens of uniform cross sections are required, a cover may be fastened to the boat; a slight tilt or a head of liquid metal is desirable in order to provide sufficient hydrostatic pressure to fill the boat. This technique provides very good control of shape.

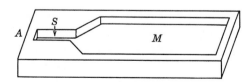

Fig. A.6. "Boat" for horizontal method for growth of single crystals.

The Czochralski method, which originated in 1917 (1), and has been widely used in the semiconductor industry, is also known as "crystal pulling." The melt is contained in a crucible, and a seed crystal that dips into it from above is rotated and slowly withdrawn, so that the rate of upward motion is equal to the rate of crystal growth, which is determined by the rate of extraction of latent heat through the seed. If the two rates do not balance, the crystal either increases or decreases in thickness as growth proceeds. The purpose of the rotation is to maintain symmetry of the crystal, the shape of which would be unstable if the crystal were stationary. An added advantage of rotation is that it causes the liquid near the interface to flow outward across the interface, decreasing the thickness of the solute-enriched layer that is present.

The fourth method, which also does not use a mold, is the *floating zone* method, in which the liquid metal is held in place by the solid above and below it, and by its own surface tension; in some cases electromagnetic forces are also used. Precise control of shape is not possible in either of the last two methods.

The "ribbon dendrite" technique, to which reference was made in Chapter 4, uses the inherent characteristics of the crystal itself to control its own geometry; the extraction of latent heat is, in this case, through the liquid, which must be maintained at a level of supercooling appropriate to the speed of growth.

A.4   Control of Composition

The compositional problems in crystal growth are twofold: to avoid changes in composition, such as contamination or the preferential loss of one component, and to produce a crystal of uniform composition.

Contamination. The problem of contamination must be divided into that caused by a solid, such as the crucible or the boat, and that caused by the atmosphere. It is not possible to state general rules to define permissible boat or mold materials, because these depend on the chemical characteristics of the substance under consideration, and on the temperature that is required. It is generally true, however, that the crucible problem becomes increasingly severe as the temperature is raised; the more reactive metals, such as titanium, zirconium, and iron, and the metals with the highest melting points, niobium, tungsten, etc., present very severe problems, and the floating zone technique, or some variation of it, is probably the only satisfactory method. Contamination by the atmosphere is more easily controlled, at least in principle, by using a vacuum or an inert gas (2).

Evaporation. Changes in composition caused by evaporation can be avoided, in suitable cases, by using a totally enclosed sample; when this is not possible, because of the lack of a suitable container material, it should be remembered that an atmosphere of an inert gas greatly reduces the evaporation rate, compared to vacuum, although it does not change the equilibrium vapor pressure of the evaporating component. It reduces the rate of evaporation by decreasing the rate at which the vapor moves away from the surface by diffusion; the layer of gas near the surface is therefore maintained closer to the equilibrium vapor pressure at which no evaporation would occur.

When a compound such as CdTe is to be prepared in the form of a single crystal, a difficulty that is often encountered is the preferential vaporization of one of the components, in this case cadmium. Techniques for overcoming this difficulty, and growing crystals very close to the stoichiometric composition, are given by Lorenz (3) who uses an appropriate pressure of cadmium vapor to suppress evaporation.

Uniformity of composition: short range. Uniformity of composition requires that there is no cellular structure and that there is no longitudinal segregation. The conditions for cell formation are given in Chapter 5, and, in principle, it is always possible to grow a crystal without cells. This usually requires extremely slow growth; however, complete suppression of the cellular structure may not be necessary since a sufficiently prolonged anneal can be used to allow diffusion to reduce to any required extent the transverse variations of composition that are a result of cell formation.

Uniformity of composition: long range. There are, in principle, two ways of avoiding longitudinal segregation. Either zone leveling or the steady state region of the "diffusion controlled" process (see page 132) may be used. Zone leveling is the name given by Pfann (4) to a group of processes in which the motion of a molten zone along the sample produces a large region of uniform composition, as distinct from zone refining, in which the composition varies continuously from one end of the sample to the other. The problem is usually to produce a uniform distribution of a solute in a solvent. If the solute is uniformly distributed in the melt, which is then solidified progressively from one end to produce a single crystal, the solute will be distributed as shown in Fig. A.7. It is assumed that the composition of the melt is always uniform, although, of course, its composition changes as solidification progresses. If, on the other hand, extra solute is introduced into a molten zone at the "beginning" of the specimen, then it will finally be distributed as shown in Fig. A.8. For low concentrations, a good approximation is obtained by putting solvent

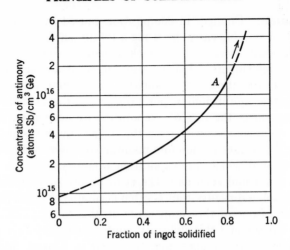

Fig. A.7. Solute distribution after normal freezing. (From Ref. 4, p. 156.)

into the starting zone only. The terminal transient, which has a high concentration of solute, can be used as the excess starting charge for the next pass, and if a zone is passed along a sample in alternating directions, the composition rapidly becomes extremely uniform.

A phenomenon referred to as "banding" has often been observed in crystals grown from the melt (5). The crystal contains bands, which demark a series of positions of the solid liquid interface, in which the composition differs from that of the surrounding material. Bands are usually, if not always, formed as a result of changes in the rate of

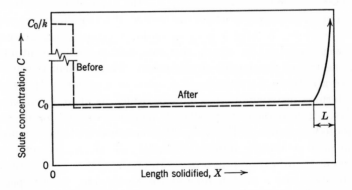

Fig. A.8. Solute distribution resulting from zone leveling using the starting charge method. (From Ref. 4, p. 154.)

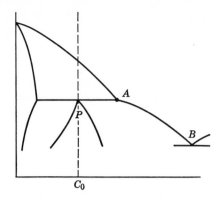

Fig. A.9. Phase diagram for peritectic system.

advance of the interface. This is sometimes caused by the operation of a temperature controlling device in the crystal growing apparatus.

Peritectic compound. A special case of composition control is encountered when a crystal is to be grown of a compound that forms by a peritectic reaction. It has been shown by Goodman (6) and by Mason and Cook (7) that zone leveling can be used for this purpose. The relevant part of a phase diagram is shown in Fig. A.9, in which P is the phase which is to be grown from the melt. The melt at the interface must be maintained between A and B, in temperature and composition. If the starting material has a mean composition C_0 (equal to that of the phase P) then it is possible to maintain a zone at a steady state composition between A and B, and a single crystal of P can be produced

A.5   Control of Perfection

A crystal grown from the melt may contain imperfections of various types, which may be classified as vacancies, dislocations, substructures and "stray crystals." The origin of these types of imperfections has already been discussed (Chapter 2). It is necessary, therefore, to show how our understanding of the origin of imperfections can be applied to the problem of growing single crystals of high perfection.

Vacancies. It has been clearly demonstrated on theoretical grounds that the vacancy content of a crystal immediately after solidification cannot be appreciably greater than the equilibrium content for that temperature. However, the equilibrium vacancy content decreases as the temperature falls, and the excess vacancies may be trapped, if cooling is fast enough to prevent the diffusion of the vacancies to sinks, where they cease to exist. This may result in the aggregation of vacancies into clusters, which can have a significant effect on the subsequent motion of dislocations, and therefore on the plastic and other properties of the crystal. This type of imperfection, which arises after solidification, can be controlled by maintaining a sufficiently low cooling rate to allow adequate diffusion of the vacancies. The same result can be achieved by a subsequent anneal followed by slow cooling.

Dislocations. The various possible ways in which dislocations may originate during solidification were discussed in Chapter 2. Success in growing single crystals with a low dislocation content depends upon simultaneously minimizing all these causes. It is, therefore, necessary to avoid the introduction of dislocations from the seed; thermal shock, either to the seed during seeding, or to the crystal when growth finishes; stresses due to the boat or mold, or to the curvature of isothermal surfaces; and local stresses due to heterogeneous distribution of solutes or to the presence of particles of a second phase. Careful attention to these aspects of the growth conditions has allowed Dash (8), for example, to grow large silicon crystals that are completely free from detectable dislocations, and Young (9) to grow copper crystals with only 5×10^3 dislocations per cm^2 compared with the more usual value of 10^7. The so-called "soft-mold" technique has been used by several investigators (10) with considerable success; a soft mold for Bridgman type growth is made by packing alumina powder round the specimen, which is then melted and subsequently solidified progressively. It has been claimed that the use of the soft mold decreases the stresses arising from differential thermal contraction of the mold and the crystal; however, the soft mold, which has a lower thermal conductivity than a boat of graphite, for example, allows the solid-liquid interface to be much more nearly planar than in the solid mold technique, and this should reduce stresses due to temperature gradients.

Howe and Elbaum (11) have shown that aluminum crystals grown by the Czochralski technique are free from detectable dislocations if they are less than 0.5 mm thick.

Lineage. It has been shown in Chapter 2 that the lineage structures form by the coalescence of dislocations, and it has been found by Elbaum (12) that, in the case of aluminum, the dislocations do not coalesce into sub-boundaries unless their density is greater than about 10^6 per square centimeter. It follows that the development of lineage structures will be prevented if the dislocation content is maintained at a sufficiently low level. It is also significant that the dis-

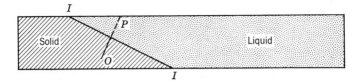

Fig. A.10. Tilted interface method for controlling dislocation content and lineage structure.

location content can be further drastically reduced by annealing if it is initially below the level at which sub-boundaries are formed.

A method that has been used with some success for controlling the dislocation content and lineage structure uses the "tilted interface," as demonstrated by Aust and Chalmers (13). This is based on the fact that the dislocations in a growing crystal tend to propagate in a direction that is normal to the solid liquid interface; Figure A.10 shows how a dislocation that originates at O will reach the surface of the crystal at P, instead of growing along the length of the crystal as in the more usual case in which the interface is either perpendicular to the length or concave to the liquid as shown in Fig. A.11. The required interface slope can be produced by a suitably disposed heater either above the boat or at one side of it.

"Stray crystals." These originate by nucleation ahead of the interface; in a pure metal, with the crystal at a lower temperature than the liquid, supercooling can exist only in regions just ahead of sharply curved portions of the interface, which Bolling and Tiller (14) have shown to exist near free surfaces and near grain boundaries (which are present in bicrystals). In practice, however, stray crystals appear to present a problem only when the melt is impure enough to permit considerable constitutional supercooling. Low speed and high temperature gradient are usually adequate to eliminate this difficulty. It is, however, observed that there are some orientations at which it is "difficult" to grow crystals, and others at which it is "easy." The latter usually correspond to the directions of dendritic growth,

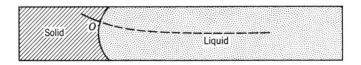

Fig. A.11. Concave interface.

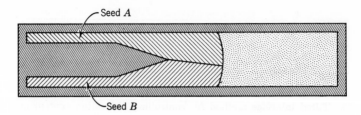

Fig. A.12. Boat for growth of a bicrystal from two seed crystals.

while the former are often the directions normal to the most closely packed planes. "Difficulty of growth" implies that stray crystals form readily, and are not easily avoided. This may indicate that the liquid ahead of the "more closely packed" surfaces is supercooled appreciably without a solute-enriched layer being present; however, other evidence does not indicate the existence of appreciable differences in the interface temperature.

A.6 Bicrystals

Many studies of the properties and the effects of grain boundaries have been based on the use of bicrystals, that is, specimens consisting of two crystals of controlled orientations, with a single boundary between them, occupying a predetermined position in the specimen. Chalmers (15) described the use of two seed crystals in a boat of the kind shown in Fig. A.12; the boundary is formed between the crystal growing from seed A and that growing from seed B. The direction of the boundary is determined partly by the orientation relationship between the crystals; but even when the orientations are symmetrical about the axis of the boat, the boundary may deviate as a result of minor fluctuations in the direction of heat flow. Davis and Fleischer (16) used projections on the bottom of the boat and on the underside of the cover to "trap" the boundary, as shown in Fig. A.13; the area

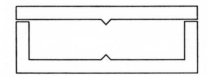

Fig. A.13. "Grooved" boat for controlling the position of the boundary in a bicrystal (cross section).

of the boundary, and therefore its energy, would increase substantially if it were to leave the preferred position. Elbaum (17) has used a similar technique for growing "tricrystals" and "quadricrystals."

References

1. J. Czochralski, *Z. Physik. Chem.*, **92,** 219 (1917).
2. For details of mold materials, etc. for specific materials, see *The Art and Science of Growing Crystals*, Ed. J. J. Gilman, John Wiley & Sons, New York, 1963.
3. M. R. Lorenz, *J.A.P.*, **33,** 3304 (1962).
4. W. G. Pfann, *Zone Melting*, John Wiley & Sons, New York, 1958, p. 153.
5. M. T. Stewart, R. Thomas, K. Wauchope, W. C. Winegard, and B. Chalmers, *Phys. Rev.*, **83,** 657 (1951). N. Albon, *J.A.P.*, **33,** 2912 (1962). H. C. Gatos, A. J. Strauss, M. C. Lavine, and T. C. Harmon, *J.A.P.*, **32,** 2057 (1961).
6. C. H. L. Goodman, *Research, London*, **7,** 168 (1954).
7. D. R. Mason and J. S. Cook, *J.A.P.*, **30,** 475 (1959).
8. W. C. Dash, *J.A.P.*, **30,** 459 (1959).
9. F. Young, *J.A.P.*, **32,** 1815 (1961).
10. T. S. Noggle and J. S. Koehler, *Acta Met.*, **3,** 260 (1955).
11. S. Howe and C. Elbaum, *J.A.P.*, **32,** 742 (1961).
12. C. Elbaum, *J.A.P.*, **31,** 1413 (1960).
13. K. T. Aust and B. Chalmers, *Can. J. Phys.*, **36,** 977 (1958).
14. G. F. Bolling and W. A. Tiller, *J.A.P.*, **31,** 1345 (1960).
15. B. Chalmers, *Proc. Roy Soc., A,* **175,** 135 (1940).
16. R. L. Fleischer and R. S. Davis, *Trans. Met. Soc. AIME,* **215,** 665 (1959).
17. C. Elbaum, *Trans. Met. Soc. AIME,* **216,** 444 (1960).

Author Index

Subject Index